Praise for Dave Eggers's

THE MONK OF MOKHA

"Exquisitely interesting. . . . This is about the human capacity to dream—here, there, everywhere." —*San Francisco Chronicle*

"A cracking tale of intrigue and bravery. . . . A gripping, triumphant adventure story." —*Los Angeles Times*

"A true account of a scrappy underdog, told in a lively, accessible style. . . . Absolutely as gripping and cinematically dramatic as any fictional cliffhanger." —*The Washington Post*

"Remarkable . . . offers hope in the age of Trump. . . . Ends as a kind of breathless thriller as Mokhtar braves militia roadblocks, kidnappings and multiple mortal dangers." —*The Guardian*

"Plainspoken but gripping. . . . Dives deep into a crisis but delivers a jolt of uplift." —*USA Today*

"A thrilling chronicle of one man's coming and going between two beloved homelands—a brilliant mirror on the global community we have become." —Marie Arana,
author of *American Chica* and *Bolívar*

"A vibrant depiction of courage and passion, interwoven with a detailed history of Yemeni coffee and a timely exploration of Muslim American identity."   —*Entertainment Weekly*

"This American coming-of-age story reminds us all of how much our country is enriched by all who call it home."   —Dalia Mogahed, coauthor of *Who Speaks for Islam?*

"Here's a story for our time: filled with ethos and pathos. You'll laugh, cry, and discover worlds unknown to most. From scamming in the Tenderloin to dodging bombs in Yemen, Mokhtar and Eggers take us on a worthwhile ride through the postmodern topography of our times."   —Hamza Hanson Yusuf

"Like many great works, Eggers's book is multifaceted. It combines, in a single moving narrative, history, politics, biography, psychology, adventure, drama, despair, hope, triumph, and the irrepressible, indomitable nature of the human spirit—at its best."

—Imam Zaid Shakir

"In telling Mokhtar's story with such clarity, honesty, and humor, Eggers allows readers to consider Yemen and Yemenis—long invisible, sidelined, or maligned in the American imagination—in their wonderful and complicated fullness."   —Alia Malek, author of *The Home That Was Our Country* and *A Country Called Amreeka*

Dave Eggers

THE MONK OF MOKHA

Dave Eggers is the author of many books, including *The Circle*; *Heroes of the Frontier*; *A Hologram for the King*, a finalist for the National Book Award; and *What Is the What*, a finalist for the National Book Critics Circle Award, and winner of France's Prix Médicis Étranger and the Dayton Literary Peace Prize. His nonfiction and journalism have appeared in *The Guardian*, *The New Yorker*, *The Best American Travel Writing*, and *The Best American Essays*. He is the founder of McSweeney's, the independent publishing company, and cofounder of 826 Valencia, a youth writing and tutoring center that has inspired similar endeavors around the world, and ScholarMatch, which connects donors with students to make college accessible. He is a member of the American Academy of Arts and Letters, and his work has been translated into forty-two languages.

www.daveeggers.net
www.internationalcongressofyouthvoices.com
www.scholarmatch.org
www.voiceofwitness.org
www.vadfoundation.org
www.mcsweeneys.net
www.826international.org

Also by Dave Eggers

FICTION

Heroes of the Frontier
Your Fathers, Where Are They? And the Prophets, Do They Live
   Forever?
The Circle
A Hologram for the King
What Is the What
How We Are Hungry
You Shall Know Our Velocity!

MEMOIR

A Heartbreaking Work of Staggering Genius

NONFICTION

Understanding the Sky
Zeitoun

AS EDITOR

Surviving Justice: America's Wrongfully Convicted and Exonerated
   (with Dr. Lola Vollen)
The Voice of Witness Reader: Ten Years of Amplifying Unheard Voices

FOR YOUNG READERS

The Wild Things
This Bridge Will Not Be Gray
Her Right Foot
The Lifters
What Can a Citizen Do?

THE MONK OF MOKHA

THE
MONK
OF
MOKHA

DAVE EGGERS

VINTAGE BOOKS
A Division of Penguin Random House LLC
New York

FIRST VINTAGE BOOKS EDITION, JANUARY 2019

The Library of Congress has cataloged the Knopf edition as follows:
Names: Eggers, Dave, author.
Title: The monk of Mokha / Dave Eggers.
Description: First edition. | New York : Alfred A. Knopf, 2018.
Identifiers: LCCN 2017032893
Subjects: LCSH: Alkhanshali, Mokhtar. | Coffee industry—California—
San Francisco. | Businesspeople—California—San Francisco—Biography. | Yemeni
Americans—California—San Francisco—Biography. | Alkhanshali, Mokhtar—
Travel—Yemen (Republic). | Coffee industry—Yemen (Republic).
Classification: LCC HD9199.U48 .E34 2018 | DDC 338.7/66393092 B—dc23
LC record available at https://lccn.loc.gov/2017032893

Vintage Books Trade Paperback ISBN: 978-1-101-97144-4
eBook ISBN: 978-1-101-94732-6

Map by Jeffrey L. Ward
Book design by Michael Collica

www.vintagebooks.com

Printed in the United States of America
10   9   8   7   6   5   4

And why? Because he let the entire world press upon him.
For instance? Well, for instance, what it means to be a man.
In a city. In a century. In transition. In a mass. Transformed
by science. Under organized power. Subject to tremendous
controls. In a condition caused by mechanization. After
the late failure of radical hopes. In a society that was no
community and devalued the person. Owing to the multiplied
power of numbers which made the self negligible. Which
spent military billions on foreign enemies but would not pay
for order at home. Which permitted savagery and barbarism in
its own great cities. At the same time, the pressure of human
millions who have discovered what concerted efforts and
thoughts can do. As megatons of water shape organisms on the
ocean floor. As tides polish stones. As winds hollow cliffs. The
beautiful supermachinery opening a new life for innumerable
mankind.

—Saul Bellow, *Herzog*

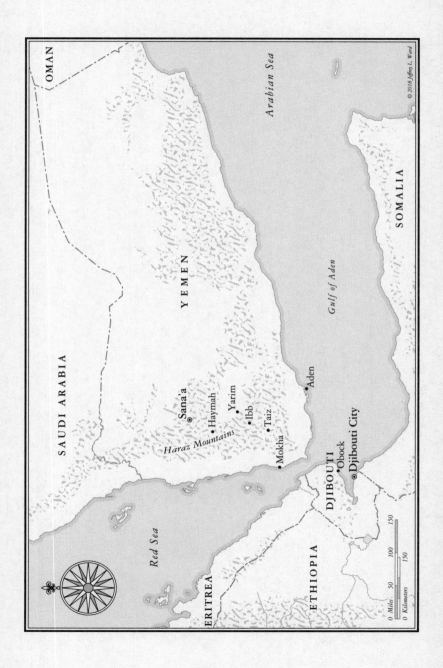

PROLOGUE

MOKHTAR ALKHANSHALI AND I agree to meet in Oakland. He has just returned from Yemen, having narrowly escaped with his life. An American citizen, Mokhtar was abandoned by his government and left to evade Saudi bombs and Houthi rebels. He had no means to leave the country. The airports had been destroyed and the roads out of the country were impassable. There were no evacuations planned, no assistance provided. The United States State Department had stranded thousands of Yemeni Americans, who were forced to devise their own means of fleeing a blitzkrieg—tens of thousands of U.S.-made bombs dropped on Yemen by the Saudi air force.

I wait for Mokhtar (pronounced MŌKH-tar) outside Blue Bottle Coffee in Jack London Square. Elsewhere in the United States, there is a trial under way in Boston, where two young brothers have been charged with setting off a series of bombs during the Boston Marathon, killing nine and wounding hundreds. High above Oakland, a police helicopter hovers, monitoring a dockworkers' strike going on at the Port of Oakland. This is 2015, fourteen years after 9/11, and seven years into the administration of President Barack Obama. As a nation we had progressed from the high paranoia of the Bush years; the active harassment of Muslim Americans had eased somewhat, but any crime perpetrated by any Muslim American fanned the flames of Islamophobia for another few months.

When Mokhtar arrives, he looks older and more self-possessed than the last time I'd seen him. The man who gets out of the car this day is wearing khakis and a purple sweater-vest. His hair is short and gelled, and his goatee is neatly trimmed. He walks with a preternatural calm, his torso barely moving as his legs carry him across the street and to our table on the sidewalk. We shake hands, and on his right hand, I see that he wears a large silver ring, spiderwebbed with detailed markings, a great ruby-red stone set into it.

He ducks into Blue Bottle to say hello to friends working inside, and to bring me a cup of coffee from Ethiopia. He insists I wait till it cools to drink it. Coffee should not be enjoyed too hot, he says; it masks the flavor, and taste buds retreat from the heat. When we're finally settled and the coffee has cooled, he begins to tell his story of entrapment and liberation in Yemen, and of how he grew up in the Tenderloin district of San Francisco—in many ways the city's most troubled neighborhood—how, while working as a doorman at a high-end apartment building downtown, he found his calling in coffee.

Mokhtar speaks quickly. He is very funny and deeply sincere, and illustrates his stories with photos he's taken on his smartphone. Sometimes he plays the music he listened to during a particular episode of his story. Sometimes he sighs. Sometimes he wonders at his existence, his good fortune, being a poor kid from the Tenderloin who now has found some significant success as a coffee importer. Sometimes he laughs, amazed that he is not dead, given he lived through a Saudi bombing of Sana'a, and was held hostage by two different factions in Yemen after the country fell to civil war. But primarily he wants to talk about coffee. To show me pictures of coffee plants and coffee farmers. To talk about the history of coffee, the overlapping tales of adventure and derring-do that brought coffee to its current status as fuel for much of the world's productivity, and a seventy-billion-dollar global commodity. The only time he slows down is when he describes the worry he caused his friends and family when he was trapped in Yemen. His large eyes well up and he pauses, staring at the photos on his phone for a moment before he can compose himself and continue.

Now, as I finish this book, it's been three years since our meeting that day in Oakland. Before embarking on this project, I was a casual coffee drinker and a great skeptic of specialty coffee. I thought it was too expensive, and that anyone who cared so much about how coffee was brewed, or where it came from, or waited in line for certain coffees made certain ways, was pretentious and a fool.

But visiting coffee farms and farmers around the world, from Costa Rica to Ethiopia, has educated me. Mokhtar educated me. We visited his family in California's Central Valley, and we picked coffee cherries in Santa Barbara—at North America's only coffee farm. We chewed qat in Harar, and in the hills above the city we walked amid some of the oldest coffee plants on earth. In retracing his steps in Djibouti, we visited a dusty and hopeless refugee camp near the coastal outpost of Obock, and I watched as Mokhtar fought to recover the passport of a young Yemeni dental student who had fled the civil war and had nothing—not even his identity. In the most remote hills of Yemen, Mokhtar and I drank sugary tea with botanists and sheiks, and heard the laments of those who had no stake in the civil war and only wanted peace.

After all this, American voters elected—or the electoral college made possible—the presidency of a man who had promised to exclude all Muslims from entering the country—"until we figure out what's going on," he said. After inauguration, he made two efforts to ban travel to the United States by citizens of seven Muslim-majority nations. On this list was Yemen, a country more misunderstood than perhaps any other. "I hope they have wifi in the camps," Mokhtar said to me after the election. It was a grim joke making the rounds in the Muslim American community, based on the presumption that Trump will, at the first opportunity—if there is a domestic terror incident propagated by a Muslim, for instance—propose the registry or even internment of Muslims in America. When he made the joke, Mokhtar was wearing a T-shirt that read MAKE COFFEE, NOT WAR.

Mokhtar's sense of humor pervades everything he does and says, and in these pages I hope to have captured it and how it informs the way he sees the world, even at its most perilous. At one point during the Yemeni civil war, Mokhtar was captured and held in prison by a militia in Aden. Because he was raised in the United States and is steeped in American pop culture, it occurred to him that one of his captors looked like the Karate Kid; when Mokhtar recounted the episode to me, he called the captor the Karate Kid and nothing else. By using this nickname, I don't mean to understate the danger Mokhtar was in, but feel it's important to reflect the outlook of a man who is uniquely difficult to rattle, and who sees most dangers as only temporary impediments to more crucial concerns—the finding, roasting and importing of Yemeni coffee, and the progress of the farmers for whom he fights. And my guess is that this captor did look like the Ralph Macchio of the early 1980s.

Mokhtar is both humble before the history he inhabits and irreverent about his place in it. But his story is an old-fashioned one. It's chiefly about the American Dream, which is very much alive and very much under threat. His story is also about coffee, and about how he tried to improve coffee production in Yemen, where coffee cultivation was first undertaken five hundred years ago. It's also about the Tenderloin neighborhood of San Francisco, a valley of desperation in a city of towering wealth, about the families that live there and struggle to live there safely and with dignity. It's about the strange preponderance of Yemenis in the liquor-store trade of California, and the unexpected history of Yemenis in the Central Valley. And how their work in California echoes their long history of farming in Yemen. And how direct trade can change the lives of farmers, giving them agency and standing. And about how Americans like Mokhtar Alkhanshali—U.S. citizens who maintain strong ties to the countries of their ancestors and who, through entrepreneurial zeal and dogged labor, create indispensable bridges between the developed and developing worlds, between nations that produce and those that consume. And how these bridgemakers exquisitely and bravely embody this nation's reason for being, a place of radical opportunity and ceaseless welcome. And how when we forget that this is central to all that is best about this country, we forget ourselves—a blended people united not by stasis and cowardice and fear, but by irrational exuberance, by global enterprise on a human scale, by the inherent rightness of pressing forward, always forward, driven by courage unfettered and unyielding.

A NOTE ABOUT THIS BOOK This book is a work of nonfiction that depicts events seen and lived by Mokhtar Alkhanshali. In researching this book, I conducted hundreds of hours of interviews with Mokhtar over the course of almost three years. Whenever possible, I was able to corroborate his memories with the help of others who were present, or with the historical record. All dialogue included in the book is as Mokhtar and other involved parties remember it. Some names have been changed. In all cases, when the dialogue takes place in Yemen, it should be assumed the language spoken was Arabic. I have done my best, with Mokhtar's help, to reflect the tone and spirit of the conversations accurately in English.

BOOK I

CHAPTER I

THE SATCHEL

MIRIAM GAVE THINGS TO Mokhtar. Usually books. She gave him *Das Kapital.* She gave him Noam Chomsky. She fed his mind. She fueled his aspirations. They dated for a year or so, but the odds were long. He was a Muslim Yemeni American, and she was half-Palestinian, half-Greek and a Christian. But she was beautiful, and fierce, and she fought harder for Mokhtar than he fought for himself. When he said he wanted to finally get his undergraduate degree and go to law school, she bought him a satchel. It was a lawyerly valise, made in Granada, painstakingly crafted from the softest leather, with brass rivets and buckles and elegant compartments within. Maybe, Miriam thought, the object would drive the dream.

Things were clicking into place, Mokhtar thought. He had finally saved enough money to enroll at City College of San Francisco and would start in the fall. After two years at City, he'd do two more at San Francisco State, then three years of law school. He'd be thirty when he finished. Not ideal, but it was a time line he could act on. For the first time in his academic life, there was something like clarity and momentum.

He needed a laptop for college, so he asked his brother Wallead for a loan. Wallead was less than a year younger—Irish twins, they called each other—but Wallead had things figured out. After years working as a doorman at a residential high-rise called the Infinity, Wallead had enrolled at the University of California, Davis. And he had enough money saved to pay for Mokhtar's laptop. Wallead charged the new MacBook Air to his credit card, and Mokhtar promised to pay back the eleven hundred dollars in installments. Mokhtar put the laptop in Miriam's satchel; it fit perfectly and looked lawyerly.

Mokhtar brought the satchel to the Somali fund-raiser. This was 2012, and he and a group of friends had organized an event in San Francisco to raise money for Somalis affected by the famine that had already taken the lives of hundreds of thousands. The benefit was during Ramadan, so everyone ate well and heard Somali American speakers talk about the plight of their countrymen. Three thousand dollars were raised, most of it in cash. Mokhtar put the money in the satchel and, wearing a suit and carrying a leather satchel containing a new laptop and a stack of dollars of every denomination, he felt like a man of action and purpose.

Because he was galvanized, and because by nature he was impulsive, he convinced one of the other organizers, Sayed Darwoush, to drive the funds an hour south, to Santa Clara, that night—immediately after the event. In Santa Clara they'd go to the mosque and give the money to a representative of Islamic Relief, the global nonprofit distributing aid in Somalia. One of the organizers asked Mokhtar to bring a large cooler full of leftover *rooh afza,* a pink Pakistani drink made with milk and rose water. "You sure you have to go tonight?" Jeremy asked. Jeremy often thought Mokhtar was taking on too much and too soon.

"I'm fine," Mokhtar said. *It has to be tonight,* he thought.

So Sayed drove, and all the way down Highway 101 they reflected on the generosity evident that night, and Mokhtar thought how good it felt to conjure an idea and see it realized. He thought, too, about what it would be like to have a law degree, to be the first of the Alkhanshalis in America with a JD. How eventually he'd graduate and represent asylum seekers, other Arab Americans with immigration issues. Maybe someday run for office.

Halfway to Santa Clara, Mokhtar was overcome with exhaustion. Getting the event together had taken weeks; now his body wanted rest. He set his head against the window. "Just closing my eyes," he said.

When he woke, they were parked in the lot of the Santa Clara mosque. Sayed shook his shoulder. "Get up," he said. Prayers were beginning in a few minutes.

Mokhtar got out of the car, half-asleep. They grabbed the *rooh afza* out of the trunk and hustled into the mosque.

It was only after prayers that Mokhtar realized he'd left the satchel outside. On the ground, next to the car. He'd left the satchel, containing the three thousand dollars and his new eleven-hundred-dollar laptop, in the parking lot, at midnight.

He ran to the car. The satchel was gone.

They searched the parking lot. Nothing.

No one in the mosque had seen anything. Mokhtar and Sayed searched all night. Mokhtar didn't sleep. Sayed went home in the morning. Mokhtar stayed in Santa Clara.

5

It made no sense to stay, but going home was impossible.

He called Jeremy. "I lost the satchel. I lost three thousand dollars and a laptop because of that damned pink milk. What do I tell people?"

Mokhtar couldn't tell the hundreds of people who had donated to Somali famine relief that their money was gone. He couldn't tell Miriam. He didn't want to think of what she'd paid for the satchel, what she would think of him—losing all that he had, all at once. He couldn't tell his parents. He couldn't tell Wallead that they'd be paying off eleven hundred dollars for a laptop Mokhtar would never use.

The second day after he lost the satchel, another friend of Mokhtar's, Ibrahim Ahmed Ibrahim, was flying to Egypt, to see what had become of the Arab Spring. Mokhtar caught a ride with him to the airport—it was halfway back to his parents' house. Ibrahim was finishing at UC Berkeley; he'd have his degree in months. He didn't know what to say to Mokhtar. *Don't worry* didn't seem sufficient. He disappeared in the security line and flew to Cairo.

Mokhtar settled into one of the black leather chairs in the atrium of the airport, and sat for hours. He watched the people go. The families leaving and coming home. The businesspeople with their portfolios and plans. In the International Terminal, a monument to movement, he sat, vibrating, going nowhere.

CHAPTER II

DOORMAN AT THE INFINITY

MOKHTAR BECAME A DOORMAN. No. Lobby Ambassador. That was the term they preferred at the Infinity. Which meant Mokhtar was a doorman. Mokhtar Alkhanshali, firstborn son of Faisal and Bushra Alkhanshali, oldest brother to Wallead, Sabah, Khaled, Afrah, Fowaz and Mohamed, grandson of Hamood al-Khanshali Zafaran al-Eshmali, lion of Ibb, scion of the al-Shanan tribe, principal branch of the Bakeel tribal confederation, was a doorman.

The Infinity was a group of four residential buildings, each with commanding views of the San Francisco Bay, of the sun-bleached city and the East Bay hills. In the Infinity towers dwelled doctors, tech millionaires, professional athletes and wealthy retirees. They all came and went through the gleaming Infinity lobby, and Mokhtar held the doors open so they could pass without undue exertion.

City College was no longer an option. After losing the satchel, Mokhtar had to get a full-time job. Omar Ghazali, a family friend, had loaned him the three thousand dollars to make the donation to Islamic Relief. But he needed to pay Omar back, and between that and the eleven hundred dollars he owed Wallead, college would have to wait indefinitely.

Wallead helped him get the doorman job; it was the same position he'd had a few years before. Wallead had been making $22 an hour, and now Mokhtar, his older brother, was making $18. When Wallead had the job, the Infinity had been unionized, but the union was gone now and the building was managed by a polished Peruvian named Maria, who clicked across the gleaming floors in high heels. She'd liked Mokhtar's clean-cut style and offered him a job. He couldn't complain, making $18 an hour when the California minimum wage was $8.25.

But he was not in college, had no clear path to college now. He spent his days in the lobby of Infinity Tower B, opening doors for residents and the various members of the service economy who kept the residents fed and massaged, the people who walked the tiny dogs, who cleaned the apartments and installed new chandeliers. Mokhtar always brought a book—he was trying *Das Kapital* again—but reading was close to impossible for a Lobby Ambassador. The interruptions were constant, the noise aggravating. The lobby was at street level, and the neighborhood was changing, a new building going up every month, turning South of Market into a kind of mini-Manhattan. The construction rattle was arrhythmic and unsettled his nerves.

The noise was one thing, but the primary impediment to getting any reading, or thinking, done was the door itself. The lobby was a glass box, a transparent hexagon, and the Lobby Ambassador had to be alert to any human coming from any angle and toward the street-facing double doors. Most of the people approaching were people he knew—residents, Infinity maintenance workers, delivery people—but there were irregular visitors, too. Guests, trainers, realtors, therapists, repairmen. Anyone coming toward that door, Mokhtar had to be ready to leap.

If it was a delivery, Mokhtar could get up, smile, open the door, no rush. But if it was a resident, Mokhtar had a second or two to leap from his seat behind the desk, rush to the door (without seeming to be desperately rushing), open it, smile and let that person in. If their hand touched the door before his, that was not good. He had to be there first, the door swinging open, his smile wide, a question ready and spoken brightly and without guile: *How was your run, Ms. Agarwal?*

All this was new. This was Maria's doing. When the building was union and Wallead was a Lobby Ambassador, the job was called a *sitting position,* meaning the Lobby Ambassador didn't have to get up every time someone went in or out. But Maria's arrival had changed that. Now the job required constant vigilance, the ability to leap up and across the lobby with elegance and alacrity.

Never mind that anyone could easily open the door themselves. That wasn't the point. The point was the personal touch. Having a smiling man in a tidy blue suit opening the door spoke of both luxury and simple consideration. It told the residents that this was a building of a certain distinction, that this well-groomed and attentive man in the lobby not only received their packages and ensured that their guests were welcomed and that unexpected visitors were vetted or thwarted, but he also cared enough to open the door for them, to say *Good morning, Good afternoon, Good evening, Looks like rain, Stay warm, Enjoy the game, Enjoy the concert, Have a nice walk.* This charming man would say hello to their dog, hello to their grandchildren, hello to their new girlfriend, hello to the guest harpist hired to play while they ate dinner.

That was a real thing. That was a real person. There was a real harpist, and he operated a company called I Left My Harp in San Francisco. Mokhtar got to know him well. For a few hundred dollars he would come with his harp and play while people ate, while people drank. A certain couple living high in the building hired him once a month. He was friendly. So was the chandelier repairman—he was Bulgarian and often stopped to talk to Mokhtar. The pet nutritionist was an affable woman with blue-streaked hair and an arm full of silver jangly jewelry. Each day a kaleidoscopic parade passed through those doors. Personal trainers, a dozen or so of those, and Mokhtar had to know them all, which among them was improving the health and longevity of which resident. There were the art consultants, the personal shoppers, the nannies, the carpenters, the concierge doctors. There were the Chinese-food people on their bikes, the pizza people in their cars, the dry-cleaning people on foot.

But primarily there were the package-delivery people. The FedEx person, the UPS person, the DHL person, bringing boxes from Zappos, Bodybuilding.com, diapers.com. Some liked to talk, some were on the clock, always late, needing just a quick signature, *thanks buddy*. Some knew Mokhtar's name, some didn't care. Some liked to chat, complain, gossip. But the volume of packages that came through that door—it was hard to believe.

What do we have today? Mokhtar would ask.

We have some cashews from Oregon, the delivery guy would say.

We have some steaks from Nebraska; these should get refrigerated soon.

We have some shirts from London.

Mokhtar signed the clipboards and brought the packages into the storeroom behind the desk, and when the resident walked through the lobby, Mokhtar raised a finger and a happy eyebrow and announced that a package had arrived. The delight was mutual. One time one of the older residents, James Blackburn, opened a box and showed Mokhtar a pair of new Montblanc pens.

Best pens in the world, Mr. Blackburn said.

Mokhtar, always polite, admired the pens, and asked a question or two about them. A few months later, at Christmastime, he found a present on his desk, and when he unwrapped it, he found the same pen. A gift from Mr. Blackburn.

For the most part the residents' money was new, and they were getting used to Infinity life. If they wanted a more formal relationship, Mokhtar could accommodate that. If they wanted to talk, he talked, and every so often there was the time and the will to have a conversation. Maybe they were in the lobby waiting for a car. Mokhtar had to be up and near the door, ready when the car arrived, so there would be those awkward few minutes, when they were both staring out into the street.

Busy today? a resident might ask.

Not so busy, Mokhtar would say. It was important never to seem flustered. A Lobby Ambassador had to project an air of calm competence.

Did you hear that new Giants pitcher moved into Tower B? the resident would say, the car would arrive, and that would be that.

But sometimes they would go deeper. With James Blackburn it went deeper. Even before the Montblanc pen, he'd shown an interest in Mokhtar. *You're a smart guy, Mokhtar. What are your plans?*

11

Mokhtar felt for him. James, a retired white man in his sixties, was a decent man, and the encounters were awkward for him, too. If he assumed Mokhtar wanted better things than working the desk and the door, that would be diminishing his current job, which for all he knew was, for Mokhtar, a personal pinnacle. On the other hand, if he assumed this *was* Mokhtar's personal pinnacle, that brought with it a more troubling set of assumptions.

Most residents didn't ask. They didn't want to know. The job, Mokhtar's existence there, was a reminder that there were those who lived in glass towers, and those who opened doors for them. Had the residents seen him reading *The Wretched of the Earth*? Maybe. He didn't hide his reading material. Had they seen him in the news, occasionally joining in or leading a protest demanding better relations between the police and San Francisco's Arab and Muslim American community? Mokhtar had been in the public eye here and there, and sometimes he thought he had a future in organizing, in representing Arabs and Muslims on some more elevated stage. City supervisor? Mayor? Some Infinity residents knew his work as a young activist, and for most he was an uncomfortable enigma. Mokhtar knew they wanted their doormen slightly more docile, slightly less interesting.

But then there was James Blackburn. *Where'd you grow up?* he'd ask. *You from out here originally?*

CHAPTER III

THE KID WHO STOLE BOOKS

MOKHTAR'S EARLIEST MEMORY OF San Francisco was of a man
defecating on a Mercedes. This was on his family's first day in the
Tenderloin. Mokhtar was eight, the oldest of what were then five kids.
For years the family had lived in Brooklyn's Bed-Stuy neighborhood,
where his father Faisal ran Mike's Candy and Grocery—a bodega
owned by Mokhtar's grandfather Hamood. But Faisal didn't want to
sell liquor, had never been comfortable selling it. After years of planning and anguished deliberation, finally Faisal and his wife Bushra
broke free. They moved to California, where Faisal had been promised
a janitorial job. He'd rather be broke and start over than be under his
father's thumb, peddling booze.

They found an apartment in the Tenderloin district, considered the city's most troubled and poor. The day they got to the city,
Mokhtar was in the backseat with his siblings when they stopped at
a traffic light. He looked over to see a white Mercedes next to them,
and just as Mokhtar was taking notice of the car, its immaculate paint
and gleaming chrome, a man in ragged clothes jumped onto its hood,

pulled down his pants and defecated. This was a block from where they were going to live.

They went from a spacious apartment in Brooklyn, from a lifestyle that Mokhtar remembered as being without want—where the kids had their own room, full of toys—to a one-bedroom apartment at 1036 Polk Street, situated between two porn stores. Mokhtar and five siblings slept in the bedroom and his parents slept in the living room. All night sirens screamed. Addicts wailed. Mokhtar's mother, Bushra, was afraid to walk alone in the neighborhood and sent Mokhtar to the store on Larkin Street for groceries. On one of his first errands, someone threw a bottle in his direction, glass crashing on the wall above his head.

Mokhtar got used to the drug dealing, which was done out in open air, all day and all night. He got used to the smells—human feces, urine, weed. To the howling of men and women and babies. He got used to stepping over needles and vomit. Older men and younger men having sex in the alley. A woman in her sixties shooting up. A homeless family panhandling. An elderly junkie standing in the middle of traffic.

The assumption in San Francisco was that the police considered the Tenderloin the city's illegal-activity containment zone—that just as the city designated Fisherman's Wharf as a quarantine for tourists, they'd designated the Tenderloin's thirty-one blocks as the city's go-zone for crack, meth, prostitution, petty crime and public defecation. Even its name, the Tenderloin, had a nefarious provenance: in the early part of the twentieth century, local police and politicians were bribed so well in the neighborhood that they ate only the finest cuts of beef.

But there was real community in the Tenderloin, too. It was one of the city's most affordable neighborhoods, and it had for decades attracted families newly arrived from Vietnam, Cambodia, Laos, the Middle East. Among them were Yemenis—a few hundred of them in the Tenderloin, most of them working as janitors. Among the patchwork legions who left their countries of origin to come to the United States, the Yemenis were late arrivals emigrating in significant numbers in the 1960s, finding work primarily in the farms of California's San Joaquin Valley and in the automotive factories of Detroit. At first almost all Yemeni immigrants were men, most from Ibb province, an agricultural region. They came to California to pick fruit, but in the 1970s, hundreds of Yemenis who had been working in the fields began to come to San Francisco to work as janitors. The pay was better and there were benefits. Eventually Yemenis made up 20 percent of the janitors' union, Local 87, headquartered in the Tenderloin.

This was Faisal's plan, too: to work in the janitorial sector, or at least start there. He got a job, but didn't last long. His supervisor, accustomed to talking down to immigrant employees—most of them from Nicaragua and China, most of them undocumented—was disrespectful. Mokhtar's father was proud and knew his rights, so he quit and got a job as a security guard at the Sequoias, a residential high-rise, on the swing shift. This was the work he did throughout Mokhtar's first years in San Francisco. His father worked odd hours, sometimes eighteen hours a day.

Which left Mokhtar free to roam. He could look in the windows of one of the adult video stores, could ignore the shirtless man screaming obscenities across the street. He could stop at one of the Yemeni markets—the Yemenis ran half the local markets, even the one called Amigo's. He could swing by Sergeant John Macauley Park, a tiny playground across from the New Century Strip Club. Up the street, on O'Farrell and Polk, there was a mural on the side of a building, an underwater scene of whales and sharks and turtles. For years, Mokhtar assumed the building was an aquarium of some kind, and only later realized it was the Mitchell Brothers O'Farrell Theatre, one of America's oldest and most notorious strip clubs—purportedly the originator of close-contact lap dances. The neighborhood had thirty-one liquor stores and few safe places for children to play, but there were thousands of kids in those desperate blocks, and they grew up quick.

By middle school, Mokhtar had become a fast learner, a fast talker, a corner-cutter, and a friend to an array of kids who also were fast talkers and corner-cutters. In the Tenderloin they dodged the junkies and hustlers and, when they could, they ventured out, knowing that a few blocks in any direction was an entirely different world. Just north was Nob Hill, one of the most expensive neighborhoods in the United States, home to the Fairmont and Mark Hopkins Hotels. A few blocks east was Union Square with its pricey shopping, the cable cars and jewelry stores.

Everywhere there were tourists, and with tourists there was always diversion. Mokhtar and his friends would go to Fisherman's Wharf and give unintelligible directions to European visitors. Or they'd ask nonsensical directions. They'd find a tourist and ask, *Do you know the way to Meow Meow? No? What about Ackakakakaka?* They'd walk by the window of any restaurant, some place they couldn't afford in their dreams, and push their naked asses against the glass. When they needed a few dollars, they'd go to the fountain in Ghirardelli Square and steal underwater coins.

Mokhtar knew his family was poor, but there were solutions to certain deprivations. He knew they couldn't afford a Nintendo 64—he'd asked for one year after year for birthdays and finally stopped bothering—but the Circuit City was only four blocks from their apartment, and that place was busy and chaotic enough that he and his friends could pretend to be potential shoppers trying out a game. Usually they could get in an hour of Mario Kart before they were chased off.

Mokhtar's neighbors were close-knit. Their building on Polk was full of Yemeni families, and they looked out for one another. The families went to the same mosque, the kids played soccer in the hallways, and for reasons beyond Mokhtar's reckoning, most of the kids were sent to school on Treasure Island. It was where a lot of Tenderloin kids went, where a lot of kids without options went. Treasure Island Middle School. It almost sounded romantic. Treasure Island itself was bizarre, an inexplicable man-made mass of contradictions. The navy built it in 1936, sinking 287,000 tons of rock and 50,000 cubic yards of topsoil into San Francisco Bay, just off of a natural island called Yerba Buena and between San Francisco and the East Bay. The island, a military base through World War II, wasn't called Treasure Island then. The name came afterward, when it was decommissioned and the powers that be, hoping to convert it to commercial use, named it after a book about murderous pirates.

But no postwar commercial anything happened, really, and the reasons were sensible but not insurmountable. First, there was some mystery about what might be buried in the landmass itself; the navy wasn't telling what kind of hazardous waste was tucked away, and no one was willing to do the research and abatement necessary. Second, there was increasing concern about where the whole island, which rested only a foot or two above sea level, would be in twenty years, given rising water levels.

At school, Mokhtar found trouble difficult to avoid. Maybe he was bent toward it. Maybe he was one of the leaders. There were Black kids, Samoan kids, Latino kids, Yemeni kids, and the boys, even at thirteen, were drinking and smoking pot, and both were done on the middle school campus—a patchwork of cement yards with narrow ranch-style buildings, each one a step up from temporary. This was the height of Mokhtar's days of cutting corners. His parents knew he was going astray. They tried to hold him accountable but he could talk himself out of any trouble. By the seventh grade they stopped listening to him.

"It's all excuses," they said.

But his teachers knew he had a mind. Mokhtar loved to read. At home, he even had a library. There was no room in the apartment for bookshelves, but on a shelf in the tiny kitchen pantry, below the canned goods and above the shelf that held the pasta and Sazón Goya seasoning, Mokhtar had carved out a home for the books he'd found. Or stolen. Getting the books involved some corner-cutting—he didn't have money to buy them, but he wanted them there, at home, lined up like they would be in a regular home. A few he borrowed, indefinitely, from the public library. His collection grew. Five books, then ten, then twenty, and soon the one shelf in the pantry really looked like something, like that one dark corner of their kitchen was some kind of legitimate haven for learning.

And because he didn't have his own room, or even his own corner of a room, the library was the one place that was his own. He collected Goosebumps books, anime, *The Chronicles of Narnia, The Lord of the Rings*. But nothing meant so much to him as Harry Potter, who lived under a staircase but didn't belong there, had in fact been chosen for great things. When Mokhtar was tired of being poor, of stepping over homeless addicts, of sleeping with six siblings in one room, his mind drifted and allowed the possibility that maybe he was like Harry, part of this hardscrabble world for now, but destined for something more.

CHAPTER IV

SAGE ADVICE FROM GHASSAN TOUKAN
PART I

THE AFTER-SCHOOL PROGRAM Mokhtar went to, at the Al-Tawheed Mosque on Sutter Street, was run by the Toukans, a Palestinian American family. Ghassan Toukan, just seven years older than Mokhtar, was one of the tutors, and Mokhtar knew he drove Ghassan nuts. Mokhtar did poorly in school and then did poorly after school. He distracted everyone. He did not care. And he did not see Ghassan Toukan, who seemed to excel naturally at everything he did, as the cure.

"Mokhtar," Ghassan implored. "Sit down. Do your homework. Do *some*thing."

Every day Ghassan hassled Mokhtar about the same things, about everything. About behaving. About homework. The wonderful advantages of completing said homework. Mokhtar couldn't take him seriously. He couldn't take any of it seriously. He was going to middle school on Treasure Island, a former military base in the middle of San Francisco Bay. It was a school for the forgotten. No one was getting out of that middle school and going anywhere that mattered.

So at the Toukans' tutoring center, Mokhtar was an agent of chaos.

He found a like-minded accomplice in a kid named Ali Shahin. Ali's father was an imam at another mosque, but Ali, like Mokhtar, was given to distraction. Together they drove Ghassan around the bend. They disrupted. They disturbed. They did no work, and the younger kids saw them doing no work, and this threw off whatever delicate academic equilibrium the Toukans were trying to engender.

"Mokhtar!" Ghassan yelled. Every day he yelled Mokhtar's name. He told him to sit, to listen, to learn.

Instead, Mokhtar and Ali snuck out of the mosque. They walked around the Tenderloin, watching out for Mokhtar's father. After years as a security guard, and after years applying for a job at MUNI, San Francisco's system of bus and tram lines, Faisal had gotten a job. He left his late-night security job at the Sequoias and now his hours were rational and steady, the benefits were good for a family of nine—he and Bushra had added two more to the brood—and the position suited his personality. He liked to drive and loved to talk.

For Mokhtar, though, his father's new job was a problem. It hemmed him in. It made him paranoid. His father's routes were different on different days, and Mokhtar could never remember where he'd be driving on any given day. So cutting corners required some care. Mokhtar and his friends would be working a hustle when one of them would look up. *Isn't that your pops, Mokhtar?* His father circled his childhood as he circled the city—a kind of sixty-foot roaming conscience.

He and Ali would go back to the mosque, back to Ghassan and his attempts to control them. And then one day Ghassan snapped. He told the four boys, Mokhtar, Ali and two other disrupters, Ahmed and Hatham, to sit down.

Ghassan pointed to Hatham. "What's your dad's job?"

"Taxi driver," Hatham said.

He pointed to Ahmed. "What's your dad do?"

"Janitor," Ahmed said.

He pointed to Mokhtar.

"Bus driver," Mokhtar said.

"Fine," Ghassan said. He knew Ali's dad was an imam, but he worried about him, too. He worried about all these kids. "Your parents came here as immigrants and they didn't have choices. Do you want to drive a taxi? Clean toilets? Drive a bus?"

Mokhtar shrugged. Ahmed and Hatham shrugged. They had no idea what they wanted to do for a living. They were only thirteen. All Mokhtar could think was that he wanted an Xbox.

"They brought you here so *you* could have choices," Ghassan said. "And you're blowing it. If you want to do something different when you grow up, you're going to have to get your shit together."

CHAPTER V

YEMEN

MOKHTAR'S PARENTS AGREED, SO they sent him to Yemen. They
thought he needed a change of location, an immersion in his ancestry,
some fresh air. Mokhtar went from his family's one-bedroom apartment in the Tenderloin to his grandfather Hamood's six-story home
in Ibb. There, Mokhtar had his own bedroom. He had his own floor.
The house had dozens of rooms, a balcony overlooking a lush valley
in the center of the city. It was a castle, really, built by Hamood from
nothing.

Hamood was more than a patriarch; in the Alkhanshali family his
influence was impossible to escape. And though he was in his late sixties, he still traveled a hundred miles a day, from Sana'a to Ibb, or out
from Ibb to the villages, attending weddings and funerals and medi-
ating tribal disputes. He was no longer a tall man—age had shrunk
him, thinned him—but his mind was quick; he was witty and tough.
Though largely retired, he was still an éminence grise in Ibb. When
he walked into a wedding hall, everyone stood. Some kissed his hand,
others kissed his head—a sign of the utmost respect.

He was born in the 1940s in Al-Dakhla, a small village inside of Ibb, the fifth of eight children. From a young age, he had the sense that he was his father's favorite. When he was still young, only nine or ten, his father was embroiled in a land dispute with another tribesman who had the favor of the ruling powers. The dispute landed him in prison, and there, his health quickly deteriorated. Knowing his end was near, he summoned only one of his children, Hamood, to his cell, and this act of favoritism soured Hamood's relations with his siblings, especially his older brothers. After their father's death, these brothers ostracized him and would not grant him any of their father's land.

At thirteen, Hamood decided to set out on his own. Without shoes and carrying only a knapsack, he left Ibb and walked to Saudi Arabia. He told this story to Mokhtar often.

"That's three hundred miles," Mokhtar would note.

"And I walked it barefoot," Hamood would insist.

Before he set off, though, Hamood asked for a donkey. He informed his brothers that he was leaving, that he would be out of their hair, and all he wanted was a donkey to accompany him, to help carry the load.

"A donkey is worth more than you," the brothers said.

So Hamood left without a donkey.

In Saudi Arabia, a land awash in oil money and far wealthier than Yemen could ever be, Hamood sold water on the side of the road. He cleaned restaurants. He did any odd job he could, and he saved money to send home to his widowed mother. And whenever he did, he enclosed a note that said, "This is from the boy worth less than a donkey."

In his late teens, Hamood went back to Yemen and married a

young woman named Zafaran, who had grown up in a neighboring Ibb village. They traveled to Sheffield, England, where Hamood had heard there was good-paying work in the steel mills. Eventually he left for Detroit, where Yemenis were finding work building cars. Hamood worked the assembly line at Chrysler, installing air bags, until, a few years later, he followed Yemeni friends to New York. With his savings, he bought a corner store in Harlem and made it profitable. He bought another in Queens, and though he had to contend with gangs and the Mafia, he was undaunted. The market in Queens did well, too, and soon Hamood was loaning money to his sons and cousins—including Faisal—all of whom opened their own grocery stores and liquor stores in New York and California, all of which paid dividends to Hamood and allowed him to more or less retire in his fifties.

He bought five acres in Ibb, and gave the builders a sketch he'd drafted himself. It was a bewildering drawing, even by the wildly eccentric architectural standards of Yemen. He wanted the house to look like the house of his imagination—the house he'd had in his mind more than fifty years, since he was a boy in Saudi Arabia. He'd just arrived there, and was still barefoot, was struggling every day to eat, when he came upon a castle on a hill. He remembered it as a castle. It might have been a hospital or mosque, but he never forgot it. He vowed that someday he would build a place like that. So he drew it from memory and when he was finished, it looked like the castle on the hill. For the interior, he did as he pleased. He adhered to no architectural precepts, no Yemeni custom. Some rooms were far larger than normal, some far smaller than normal. Some floors had four bathrooms when none at all were needed. There were balconies everywhere, but the entrances and exits were never in the expected places. *If a thief ever tries to steal anything from that house,* Zafaran said, *they'll get lost and never get out.*

He began building the house in 1991 and never finished. When Mokhtar arrived and throughout the year he spent in Yemen, there were always workers in the house. At any given time there were five craftsmen adding custom touches, all according to Hamood's specifications: a new door carved from rare teak, imported tile in the fifth-floor sitting room, new stained glass above the fourth-floor balcony. Walls were covered with his collection of daggers, swords, cowboy hats, holsters and guns. He had a Beretta, an array of Colt .45s, a collection of pistols he'd seen in Bond films and John Wayne movies. Hamood had seen every film John Wayne had ever made, and collected holsters, hats, wore cowboy boots—anything Wayne had worn, he wanted.

When Mokhtar got to Ibb, just after eighth grade, he had no interest in John Wayne and no interest in Yemen. He missed the action in San Francisco. Hamood sent him to a local school, private and rigorous, and made him walk to it, forty-five minutes each way. Mokhtar spoke some Arabic, but no one in the school spoke English. He was one of the only Americans there. He didn't wear his clothes correctly. He didn't know the proper responses to standard greetings. He didn't know the right Yemeni way to walk, act, smile, not smile. To fit in, he decided he would become super-Yemeni. He worked on his Arabic, ground down his accent, dressed like the Yemeni kids, with a sarong and sandals and the right kind of jacket. He tried to assimilate and master local customs, but the embarrassments were unending.

One day Mokhtar's grandmother Zafaran sent him to get a chicken. Mokhtar was used to American groceries, where the chicken had been processed hundreds of miles away, sliced up and wrapped in plastic, unrecognizable as a once-living thing. Now he had to ask the butcher in downtown Ibb for a chicken, which he did successfully enough. The butcher reached for a live chicken and asked Mokhtar another question, which he didn't understand, but felt it best to answer in the affirmative. The butcher was surprised but shrugged, took the chicken, cut off its head, and put it in a plastic bag, still bleeding prodigiously and covered with feathers.

When Mokhtar got back to Hamood and Zafaran's house, holding a plastic bag heavy with chicken blood, Zafaran stared at him. Then she laughed. Mokhtar had a feeling she would call him Dummy, because she often called him Dummy—she called everyone Dummy; it was her favorite word in English.

"Dummy," she said.

Mokhtar left the chicken in the kitchen with one of the maids and went to the living room, where Hamood was hosting guests. There were always guests for lunch, no invitation needed. Mokhtar was enjoying his lunch with these neighbors when Zafaran burst in and told the story of the chicken.

"Dummy," she said. "What a dummy!" Everyone laughed.

But soon Zafaran and Hamood began to trust him with tasks both small and significant. "Go to the bank and cash this check," Hamood would say, and would hand him a check for three million riyals—about fifteen thousand dollars. Mokhtar would return, navigating the streets of Ibb carrying an enormous bag of money like a cartoon bank robber. Hamood had business all over Ibb, and all over

Yemen. He brought Mokhtar on his rounds, teaching him how a businessman carried himself, how a leader walked and talked. The tasks Hamood concocted for him were far-flung and grand. One time he gave Mokhtar a bundle of cash and instructed him to go to Taiz, two hours away, and come back with six tons of a certain kind of stone he needed for the courtyard of the house. Mokhtar returned that evening, leading a caravan of three full flatbed trucks.

When Mokhtar made a mistake, Hamood was angry only if Mokhtar made an excuse. "Own the error and correct it," he said. Hamood had a thousand proverbs and maxims. His favorite was *Keep the money in your hand, never in your heart.* He used that one a lot.

"What does that mean?" Mokhtar asked.

"It means that money is ephemeral, moving from person to person," Hamood said. "It's a tool. Don't let it get into your heart or your soul."

Mokhtar spent a year with Hamood and Zafaran and returned to the United States changed. Not entirely reformed—there was still significant corner-cutting in high school—but he'd studied classical Arabic, awakened to his Yemeni heritage, and though Hamood hoped Mokhtar might become an imam or an attorney, Mokhtar began instead to see himself in Hamood's mold, as a man of enterprise. A man who liked to move.

CHAPTER VI

RUPERT, ARRIVISTE

NOT LONG AFTER RETURNING from Yemen, Mokhtar was wearing sweater-vests and working at Banana Republic. His Tenderloin friends were confused. His friends from middle school, startled by the transformation, began calling him Rupert, like the spiffy animated bear. Mokhtar didn't mind. He was fifteen and proud to have a job. When he got back from Ibb, he wanted a job, saw a listing at Banana Republic, applied, and got it.

His parents couldn't believe it. No one believed it. A Tenderloin kid working at Banana Republic. And not some backwater Banana Republic at the mall, but at the flagship store downtown. When he got the job, he'd expected to be put in the back room, and that's where he started, but soon he was on the main floor, selling shirts and khakis to businessmen and tourists.

For Mokhtar, it was a time of radical evolution. He met his first gay man. Mokhtar had spent years in San Francisco and never met a gay dude. Or maybe he had—he probably had—but he hadn't known it. Not out gay guys like his bosses and coworkers at Banana Republic. They welcomed him, they taught him what went with what, how to fold a marbled cable-knit cardigan, how to hang a pair of Kentfield slim cotton pants. He spent most of his paycheck on clothes—vintage woven-trim henleys, $130 leather shoes, English-cut pants that ended high at the ankle.

The effect of his appearance on the world was profound. He walked through the city not as a poor Tenderloin kid, his baggy clothes shrouding him with negative assumptions, but as Rupert, the preppy cartoon bear, who was welcome anywhere. He became someone the adults he encountered—at his high school, at the mosque, at any store he entered—trusted and wanted around.

He was called *sir*. He was called *mister*.

After a year at Banana Republic, he heard about a job at the Union Square Macy's, selling shoes to women, and even though he was seventeen and knew nothing about women's shoes, or about women, or Macy's, he applied for a job, and in his Rupert incarnation, he got the job. The commissions were higher than Banana Republic's, so he left that gig for Macy's, and on his first day, holding his breath, he held the trembling foot of a thirtysomething woman in a short skirt.

I would heartily recommend this job, he told his friends.

You weren't supposed to ask out the customers, and he didn't have to. They pursued him. Every day there he was, well dressed and kneeling before them, holding their unslippered feet. They were Cinderella, he was Cinderella. He was the interloper to the ball. He didn't know their world. A pair of women would come in, holding Gucci purses, fondling the shoes, talking about vacations in Madrid and Cannes and Saint Bart's. On Monday he'd hear a woman telling her friend about her son who wanted to go to USC, how they had an excellent film school, and on Tuesday, hearing another mother talking about her own creative son, he'd expound with great authority about how good and selective the USC film program was. *Probably the best in the country,* he'd say.

The Tenderloin taught you to think quick, talk fast. You had to listen and assimilate. If you sounded ignorant, you got taken. Within a day or two at Macy's, he knew Cole Haan, Betsey Johnson, Coach, Vince Camuto, Michael Kors, and was making about two hundred dollars in commissions a day. He averaged twenty hours a week, after school and on weekends, and there were women who believed, or let themselves believe, that Mokhtar was older than he was. There were the twentysomething sisters from Germany. There were the thirty-something women from New York. He and another shoe guy would take them out, or let themselves be taken out, show them spots in the city they wouldn't otherwise know. Nothing much came of any of these dates, but he learned. He learned what it was like to travel, to have the money to buy things, to buy plane tickets to the Caribbean, to Europe. When you're in Paris, these women would say, you have to go to L'Abeille! And don't go to Jackson Hole in January. December or February, but never January. Good to know, he told them, and every night, he went back home, to sleep on the top bunk of a two-bunk set in his family's one-bedroom apartment on Polk.

By eighteen, he knew these people, who had gone to college and could live wherever they wanted, had nothing he didn't have. They weren't any smarter, this was clear. They weren't quicker. They weren't even more ruthless. If anything, they were softer. But they had advantages. Or they had expectations. Or assumptions. It was assumed they'd go to college. It was assumed they'd find jobs befitting their upbringing and education. There were no such assumptions in Mokhtar's world. In high school there had been the odd teacher who mentioned college to him, saying he could do it, he had the mind, but there wasn't much college talk at home. There was no precedent and there was no money.

CHAPTER VII

RUPERT SELLS HONDAS

A FEW MONTHS AFTER his high school graduation, Mokhtar saw a help-wanted ad for a valet. Honda of San Francisco, a dealership on Van Ness, was looking for someone to park cars.

Mokhtar filled out an application and found himself sitting across from a sturdily built man named Michael Li. Quickly Mokhtar learned that Li had been a marine, did time in the First Gulf War and now ran the sales floor at the dealership. He asked Mokhtar questions about cars, about his work experience, and Mokhtar told him about Banana Republic, about Macy's, and exaggerated a bit about his own car knowledge. He had two uncles downstate near Bakersfield, Rafik and Rakan, who had taught him a few things about cars. Mokhtar threw out some terms—*alternator, dual-quad, carburetor*. Li nodded, listened, asked questions. The job interview went on longer than an interview for a car parker should, and finally Li came out with it: "You ever think about selling cars?" They'd just lost a sales associate, he said. Would Mokhtar want to give it a try?

Mokhtar was prepared. He was always prepared. Any Tenderloin kid was always prepared. His mind worked quickly enough that after a few minutes with Li he'd already sensed some possibility in the air, and while he was answering questions about folding shirts and selling shoes, the parking of cars, another part of Mokhtar's mind was assessing the likelihood that the marine would offer him some other kind of job. Mokhtar couldn't tell people about this kind of thing—how he could sniff some opportunity and mentally prepare himself for it. They didn't understand. But he knew that if anyone gave him the slightest opening, if the door were even an inch ajar, he could talk himself all the way in.

Which is what he did with Li. Sure, he'd thought about selling cars, he said. He went into high-bullshit mode. He'd thought about it a lot, actually. Especially Hondas, Mokhtar said. Hondas are so reliable. And the resale value! He looked out to the lot and threw out random-but-general-enough-to-be-accurate thoughts about the Accord, the Civic, that bizarre box-on-wheels called the Element. He talked about ABC, Always Be Closing. Where had he heard that? He used it, it felt right, and Li continued to nod. After half an hour of hearing Mokhtar talking out of his ass, Li hired him as a junior sales associate.

Mokhtar was nineteen years old.

He brought a dozen brochures home, studied the various models and features, and came back feeling invincible. Now that Mokhtar was working there, Li shifted into a different man—not the guy who had interviewed him. That guy had been so gentle. He'd spoken with a small and delicate voice, strange coming out of his square jaw and thick neck. But that was Li's way during interviews, and with customers, too—his voice tiptoed around, his smile was kind, his posture relaxed. But off hours, behind closed doors, talking about quotas, moving inventory, Li was a marine. *You gotta control the motherfucking conversation, Mo! Don't let those fuckers get control. Don't let those fuckers get control! Whoever controls the conversation controls the deal, you got that?*

Mokhtar couldn't argue. The guy was about a buck-ninety and cut like a statue. So Mokhtar tried to control the conversation. Ask them questions. Questions they have to say yes to, Li said. *Get 'em saying yes. Get them fucking saying yes, you understand?*

Mokhtar understood. He'd see a prospect on the lot, a middle-aged guy wearing a 49ers hat. He'd saunter up.

You like the 49ers this year?

Sure.

That Justin Smith, he's a beast, right?

He is.

And Frank Gore! That guy's a tank! Plays the game how it should be played.

Right.

(Now look at the car, a black Accord. Keep the yeses coming.)

You like this car?

Sure.

You like this color?

Yeah.

Can't beat black. Always looks good, day or night. You want to get inside?

Sure.

You like the dash?

Yeah.

You like the leather?

I do.

Check out the digital speedometer. You like that?

I do.

Check out this sound system. You like Tupac? Coldplay?

Coldplay.

Me too. You see that concert last year? At Shoreline? Oh, and check out the GPS. You like it?

Yeah.

You want to take it for a drive?

Sure.

You like the acceleration?

Yeah.

You like the steering?

Yeah.

The cornering?

Sure.

The digital speedometer?

Yeah.

If we get the right price for you, you think you'd want to drive this baby home today?

After the test drive, it was up to Li. That was the arrangement. Li had decided that Mokhtar would be the young car guy, the young upstart who knew cars, loved cars, but didn't know prices. The numbers weren't his forte. So Mokhtar would grab a prospect, get them excited about the car, just one car lover to another—doing the test drive, cranking some jams, cruising around South of Market, sweet. Then he'd bring them back to the office and Li would step in.

The second month Mokhtar sold two cars. The third month, nine. Soon Li let him deal with the numbers, make the offers. First

Mokhtar had to learn how to size people up. He knew clothes, knew when someone could afford a nice shirt, nice shoes. The shoes were key, but sometimes deceptive. The tech people all wore sneakers, and sneakers had a low ceiling. But he learned how it worked. Some of the wealthiest people liked simple cars and paid cash. The aspirational types wanted the car loaded with extras, and they liked to finance. Either way the price could be massaged. There were four boxes—total price, interest rate, monthly payments and down payment—and you could toggle each one till you got the price you needed. But first was the offer, the base number, and how that number was delivered was everything.

Make an offer and shut up, Li said. *You make the offer and whoever speaks first loses, you got that? Whoever fucking talks next loses.*

Mokhtar would say a number, $32,500, and stare at the customer sitting on the other side of the desk. Just stare. Nothing bizarre—he wasn't trying to hypnotize anyone. But you had to be confident in that number. The number was the number. It was the best number you could do. And always the customer would speak first. Always. *You let that fucker talk first, you hear me? Whoever fucking talks first loses.*

After a while Mokhtar was averaging twelve cars a month and was pulling down three thousand dollars a month in commissions. He bought himself new clothes and new shoes. He gave the rest to his parents. They were proud and saw no reason for him to deviate from that course. Selling cars—it was more than he'd make working behind the counter at any of the Yemeni corner stores. More than he'd make pushing a mop.

But after a year, Mokhtar felt an itch. He wanted something else, something more. A few of his friends were going to college, or were already in college, and he was thinking of making a move. Maybe he was just looking for an excuse.

There was an old man on the lot. Impossibly old. Mokhtar couldn't understand how the old man had driven his way to the lot, had gotten out and was walking around. He looked at least ninety. Mokhtar made his way over to him, and the closer he got, the older the man seemed. A hundred and ten, easy. He was dressed in an old-man way and he wanted a new car. He wanted to trade in his Chevy station wagon for an Accord, he said, but he didn't like the Accord's sticker price. Mokhtar had been told about these men, the ones who had read some *Consumer Reports* article, or something on the Internet, and came to the dealership with some arbitrary price in mind. This price was never the sticker price. It was always some number below the sticker price. Sometimes it was five hundred dollars under. Sometimes it was fifteen hundred. Usually it was one thousand. This was the standard number. They wanted a thousand off whatever the sticker price said. And this old man was no different. He told Mokhtar he wanted the car, but wanted a grand knocked off. Mokhtar told the old man he'd check with his manager, and he went inside to see Li.

Sure, Li said. *Let's get the paperwork written up.*

Mokhtar went back outside and told the old man that they had a deal. They'd knock a thousand off the sticker price.

The man was elated, and they shook hands, and Mokhtar brought him inside and introduced him to Li. *I'll take it from here,* Li said, and Mokhtar went back to the lot.

An hour later Mokhtar watched the old man drive off in his new Accord, and he waved to the old man, thinking that something good had been done that day. He was impressed by Li, by the way a brutal negotiator could show a certain respect, a certain mercy, for an old man, call a cease-fire that day in the endless commission skirmishes, and go ahead and just knock the thousand off. The old man only had so much time left on the planet; the hassle wasn't worth it.

That was cool of you, Mokhtar told Li.

Li looked at Mokhtar funny, and pointed to the contract. The numbers were the usual garble of add-ons and fees and other nonsense, and Mokhtar realized Li hadn't knocked off anything. The car cost the old man exactly what Li wanted to charge him. He'd taken $1,000 off the initial number, but he'd added it right back on.

That made it easier to leave. Mokhtar had been part of a hundred deals before, and there was always some number juggling, but this was different. Mokhtar went home that day, and a few days later he resigned, via text. He knew it was unprofessional, and that it spoke to the declining standards of propriety and workplace decorum, but he was done. That's what he wrote to Li in his text, just three words: *I'm done, son.*

CHAPTER VIII

RICHGROVE AGONISTES

BAKERSFIELD IS NOT HIGH on the list of places a young man might go to begin his hero's journey. But Mokhtar's grandmother Sitr, his mother's mother, lived in nearby Richgrove, and she had a couch. Mokhtar planned to sleep on this couch, rent-free, while he took classes at Bakersfield College. He needed time to concentrate on school without spending money, so he took a bus four hours downstate.

It was not entirely his choice. His father Faisal was not happy about him quitting the Honda job. The way Faisal saw it, Mokhtar was well paid there, had a future there, and he quit for no significant reason. It was further evidence of a restless, or even shiftless, disposition. Mokhtar's parents wanted to know that Mokhtar had a plan. Either get and keep a job, they said, or go to college. Quitting good jobs while continuing to sleep on their floor was not a tenable option.

Live down here for a while, his grandmother said. *Take some classes.*

Mokhtar needed some air, a change of pace. He packed his stuff and enrolled in four courses: political science, world history, sociology

40

and film studies. He moved into Sitr's house, just behind the Fastway Gas and Grocery on 99 South.

Sitr and her husband Ali had purchased the Fastway back in the 1980s. The business was in the middle of the Central Valley, surrounded by fruit farms—grapes across the highway, avocados and almonds down the road. When Sitr and Ali took over, it was the only gas station for miles, and it was instantly profitable. For the fruit pickers, almost all of them from Mexico and elsewhere in Latin America, Fastway was the place they'd buy lunch, buy beer for after work, gas for their trucks. The fuel never made a profit—no gas station can turn much of a profit on actual gas—but it brought people into the grocery, and that's where the profits happened. Food, lottery tickets, liquor.

Ali and Sitr built a compound behind the station and lived there, happily, for twenty-five years. Their kids and now grandkids were brought up working at the store, speaking Arabic at home, English in school, Spanish in the grocery. Mokhtar had been going downstate to visit since his family had moved to California. It was a rural life—quiet and dry and hot. On the narrow roads cut between fields of plums and grapes, his uncles Rakan and Rafik had taught him to drive. And because in Yemen, basic facility with a rifle was expected of any young man, they took him to the 5 Dogs Range to teach him how to shoot.

When Mokhtar arrived this time, his grandfather Ali had been dead ten years, but the business was still thriving, still owned by Sitr, and run by Rakan and her son-in-law Taj. Taj and his wife Andrea lived with their four kids in the compound, too. It was crowded, but they made room for Mokhtar. He slept on the couch for the first month or so, until Rakan found and resurrected an old bed frame from the garage; it had been used, years before, by Andrea's daughter Khitam. It was pink, but he was grateful, and he tried to be useful at the Fastway. He took out the garbage. He broke down cardboard boxes. He helped out Olga, the sharp-tongued Mexican American cook who made burritos and empanadas and sandwiches for the farmworkers.

The laborers who came to the Fastway would give Sitr boxes of grapes, oranges, plums, blueberries and almonds—whatever was in season—and Sitr would give them the herbs, spices and figs she grew in the compound. Sitr loved Richgrove. It reminded her of Yemen—the warm and fertile Yemen she'd known as a girl. Everything grew in Ibb, she told Mokhtar. Melons, figs, lemons, apples, almonds. Anything. Their Yemeni ancestors had been farmers in their native province of Ibb, so it made sense that Yemenis from Ibb came to California. It was where she was meant to live.

Sitr had known Cesar Chavez and Dolores Huerta. She'd known Nagi Daifullah, a Yemeni American farmworker and a martyr to the farmworkers' cause. When Cesar Chavez began to organize farmworkers, the Yemenis in the Central Valley lined up behind him. In 1973, Daifullah, a Yemeni from Ibb, became a United Farm Workers' strike captain. Fluent in English and Spanish, he was a crucial link between the Spanish- and Arabic-speaking laborers. In August that year, at the height of the UFW battles with the farm owners and law enforcement, Daifullah was outside a bar, celebrating a modest union victory. A Kern County police officer approached, and he and Daifullah had words. He beat Daifullah over the head with a flashlight and dragged him through the street, killing him. Chavez himself led the funeral procession, seven thousand farmworkers strong, through Delano.

Now Mokhtar was going to college with the sons and daughters of these farmworkers. His classes at Bakersfield College were decent but he was soon bored. To some residents of Richgrove, Bakersfield was the big city, but Mokhtar found nothing to do in Bakersfield. He had no car and no spending money. He didn't connect with the other students. There was a Persian woman in his sociology class he found intriguing, and there were a few Muslim women who found him intriguing, but otherwise he quickly realized Bakersfield was not his destiny. After a semester he was gone. He moved back in with his parents, who now lived on Treasure Island, a few blocks from where he'd gone to middle school.

They were unhappy with him. He'd quit his Honda job. He'd quit college. Now he was sleeping on their floor.

But there was hope in Yemen. It was 2011, and Yemen was swept up in the catapulting hopes of the Arab Spring, and Mokhtar joined the Yemeni American community in the Bay Area to celebrate the progress made and to try to articulate the possibilities at hand. In April, Mokhtar and a group of other young Yemeni Americans organized a march, and two thousand Yemeni Americans demonstrated in San Francisco to support Yemenis' push for democratic change. Shortly after, he was part of a national delegation of Yemeni Americans invited to Washington, D.C., to address the State Department and the White House. Mokhtar was twenty-one years old, the youngest of the delegation, comprising nineteen representatives from eleven states, and he had nothing to wear. He had owned one suit in his life, and he'd worn it through.

Mohamed Mugali, the imam from the 7th Street Mosque in Oakland, brought him to Men's Wearhouse and hooked him up. Mokhtar had no funds for the flight to D.C., so Mugali and a group of activists paid for that, too. Mokhtar showed up at the San Jose Airport at 6:00 a.m., only to find that Mugali, a novice at online ticketing, had booked Mokhtar's ticket from San Jose, Costa Rica, not San Jose, California.

The airline took pity on him and he was in D.C. that night. The next day the delegation—they called themselves Yemenis for Change—addressed the State Department, outlining what they saw as the traditional pair of choices for Arab countries in the Middle East, military dictatorship (Libya, Iraq, Egypt) or right-wing theocracy (Iran, Saudi Arabia).

Their presentation was called the Third Option, and pointed to Cairo's Tahrir Square, to the tens of thousands of young Egyptian activists who wanted democracy, who harbored no ill will toward the West, who wanted a self-determining nation based on a constitution—a new constitution—and the rule of law.

The State Department people were politely intrigued. Then the delegation made a request. The United States had to stop supporting Ali Abdullah Saleh, the president of Yemen, to whom they were funneling $200 million in weapons that year.

The State Department people were nonplussed, but the delegation was invited to the White House, where they gave pretty much the same presentation to a small group of President Obama's foreign policy advisors. The results were unclear, but the delegation left Pennsylvania Avenue feeling heard and gratified, and Mokhtar and two other delegation members, Mugali and Hesham Hussein, a chemical engineer from California, went on to the Lincoln Memorial.

"See what you can have in Yemen?" Hussein said. He was standing at the foot of the monument, talking into a video camera—he was making a video message for the protesters in Sana'a. He hoped the video would inspire them. "Wouldn't you like this kind of freedom in Yemen?" he asked the camera, explaining the work of the delegation, that they'd had audiences at the State Department and White House that day.

Mokhtar was feeling well pleased. The United States could make terrible mistakes abroad, and particularly in the Middle East, and the matter of the drone strikes was one that the delegation couldn't agree on how to address, but at the same time there was a certain openness, the ability to say anything you wanted—that was real, and as an American, he felt proud of it. And then, out of the corner of his eye, he saw a uniformed man moving toward them. The man was wearing blue, and had a badge. *Please no,* Mokhtar thought.

"Excuse me," the man said. He was pink-faced and smiling. "Hi there, how you guys doing?"

Mokhtar, using his most American English, said they were doing fine. He had a sense where this was headed but prayed he was wrong.

"What, uh—what language were you guys speaking there?" the officer asked. Mokhtar looked closely at his badge. He wasn't D.C. police. He was something else. Not Secret Service, but some kind of police force dedicated to the monuments.

Mokhtar told him they had been speaking Arabic.

"Arabic, huh?" the officer said, and his eyes seemed to register, for a moment, that he was onto something potentially serious. "You mind if I see your IDs?"

By now Hussein had stopped videotaping. They handed the officer their IDs, and the officer jogged down the steps of the Lincoln Memorial to a black car. Mokhtar assumed he was running their names through a database of terror suspects. By now, most of the people visiting the monument were watching the events, stealing glances at Mokhtar's trio. Some tourists had made a quick exit—possibly thinking some violence was about to erupt between law enforcement and a group of extremists.

Mokhtar thought of his father, who was already on some kind of registry, he was sure. Just a few years earlier, Faisal and Bushra had been driving through Treasure Island, looking at potential homes to rent, when they were pulled over by police. Someone had seen them driving around the island, Bushra wearing a hijab, and thought they might be casing the area for a potential terrorist strike. Faisal and Bushra had received a kind of apology eventually, but Mokhtar had no doubt that their names, and perhaps his, were on some shadowy database, never to be purged.

Fifteen minutes later the officer returned to the feet of Lincoln.

"Sorry about that," he said. "You're free to go. Or stay."

Now, though Mokhtar knew it wasn't a good idea to escalate any of this, or even extend it a minute into the unknown, he couldn't help it.

"Officer," he said, "what if I told you that I'm an American citizen, and that we just came back from the State Department and the White House, where we were asked to speak? And after a day speaking to important people and feeling good about our democracy, now this

46

will be my experience in D.C.? Because that's what just happened. If Lincoln were alive, what would he say?" Mokhtar went on this way for a while, until the officer's face seemed to soften. His eyes weren't the eyes of a zealot or an ignorant man. They were the eyes of a man acting on orders and with limited information.

"Well, I'm sorry," the man said. He apologized a few more times, and seemed to actually mean it. He jogged down the steps, and back into the car, and the car rolled away.

CHAPTER IX

THE BUTTON

FOR THE NEXT FEW years, Mokhtar had no plan. He slept on the
floor of his parents' Treasure Island home, and worked temp jobs. He
spent time at UC Berkeley, helping organize students around causes
crucial to Arab and Muslim Americans. He spent so much time there
that most of the students, Ibrahim Ahmed Ibrahim among them,
assumed he was enrolled. But he wasn't taking classes there, or anywhere. He watched his peers become sophomores, juniors, seniors. He
watched them graduate. He watched Miriam graduate. He lost years
to indecision, to inaction.

For a while he worked for Omar Ghazali, a prosperous fruit broker. Omar had grown up in Yemen, and came to the United States in
2004 with nothing and with no particular plan in mind. He drove
a taxi for a while, then worked as a security guard, then a valet, and
finally tried his hand at buying and distributing California produce.
He bought from the Central Valley and sold to San Francisco. Soon,
most of the fruit available in Chinatown came through him. The Mission, too. If someone needed ten thousand oranges by the next afternoon, he could get them. Ten tons of Stockton cherries overnight—he

could make that happen. He grew a tiny business into a multimillion-dollar operation.

He gave Mokhtar a job in the Oakland warehouse, loading trucks. Sometimes Mokhtar made deliveries. He called delinquent accounts. He learned that the best California cherries were exported to Japan and could bring as much as one dollar for each cherry. He learned that it mattered which farm produce came from—that an orange from Stockton tasted different from an orange from downstate. He learned, too, that Omar didn't really need him. He'd given him a job as a favor to a fellow Yemeni, and when Mokhtar had saved enough to pay for classes at City College, Mokhtar was free to go.

Mokhtar took his savings and enrolled. Then Miriam gave him the satchel. And he bought the laptop with money borrowed from Wallead. He raised money for Somali famine relief. Then he lost it. Omar loaned him the money to give to Islamic Relief, and now Mokhtar was forty-one hundred dollars in debt.

So he was a doorman, and every day he sat at his Infinity desk, vibrating. Thinking of time slipping. Friends were going to graduate school. His younger brother Wallead was about to graduate from UC Davis. Mokhtar was twenty-five and had four community-college credits to his name.

He was a doorman. A doorman listening to the inanities and vulgarities of the Infinity residents. A woman had recently spent fifteen minutes on her phone, in the lobby, engaged in a lewd sexual conversation. She knew he could hear; she was no more than five feet away. She didn't care, or found it entertaining, or alluring, to talk graphically in front of him. Was she worse or better than the resident who made a point to tell him about the eighty thousand dollars' worth of china she'd had delivered? Why did he need to know that? At the holidays, she gave him twenty dollars and a cookie.

But he was thankful for the paycheck. Thankful to be working in a clean and safe place, in a job that was not difficult or dangerous. He had friends in jail. He had friends working in Tenderloin corner stores, shotguns within reach. And Ali Shahin, the kid from the Toukans' tutoring program, the son of the imam, was dead. He'd gone to Mecca, and within weeks of returning to San Francisco, he was found near Candlestick Park, shot five times in the head. No one knew who did it or why.

Mokhtar sat at his Infinity desk, knowing that could have been him. He and Ali knew all the same people. Had seen the same things, were seduced by the same temptations. Mokhtar, alive and safe in the Infinity, felt grateful. But he wanted more. He just didn't know what.

Justin wanted to be an olive importer. Justin Chen was a friend Mokhtar had made at UC Berkeley—one of the many students there who assumed Mokhtar was enrolled. Every so often Justin would come by and sit on the Infinity lobby's white leather couch. Maria didn't allow Lobby Ambassadors to visit with friends, but Justin could pass for a bike messenger, and he and Mokhtar could kill half an hour, Mokhtar hovering between the desk and the front door in his blue suit, opening the Infinity doors while Justin talked about olive oil.

Justin was finishing up his degree in peace and conflict studies, but what he really wanted to do was grow olives. Mokhtar listened, half-amused and half-exasperated. What did Justin know about olive oil? Justin wanted to buy land in California, grow olives, package olive oil. *Specialty* olive oil, he said. He'd studied the supply chain and had ideas for improvements. Mokhtar didn't know what to say. Justin had no family in farming in California. Why olives? Hadn't he wanted to be a cop at one point? And where would he get the money for an olive farm?

Sometimes Miriam came by. Miriam had finished college and was helping out at her parents' business, Ted's Market at Howard and 11th. Sometimes she made deliveries for the deli, and when the deliveries were at or near the Infinity, she'd stay with Mokhtar and pass the time until Maria came around.

Their romance lasted a year, maybe less. There were the obvious obstacles. Mokhtar was from a conservative Yemeni family—the Yemenis being the most insular of Arab communities. It was almost unheard of for a Yemeni American to marry outside the community. Most of Mokhtar's Yemeni friends, male and female, had entered into arranged unions with Yemenis from the homeland. That was the standard—you go back to Yemen, marry whomever your parents set you up with, natives of Ibb or Sana'a or Aden, the two families going back centuries together. Rare were the instances of Yemeni Americans meeting and marrying other Yemeni Americans, and unheard of was any Yemeni American man marrying a woman whose mother was Palestinian and whose father was a Greek American Jerry Garcia devotee—and both Christian. It was impossible.

So Mokhtar and Miriam had been careful. They went into it slowly, chastely, always on the lookout for Mokhtar's dad, circling the city in his bus. When, after weeks of flirtation, they finally admitted to each other that their feelings were romantic, they spent all night

51

walking the city, and at last made their way to Ocean Beach, where Mokhtar had long wanted to take her. The night was clear, the sand was warm from a day of sun, and all was good until it was 3:00 a.m. and they were waiting for the bus home. As it approached, Mokhtar remembered—how could he have forgotten?—that they were on his father's line, the 5 Fulton, and if he caught them together, there would be formidable woe. So they ran from the bus stop and walked the many miles till she was home.

Now their friendship was more important to Mokhtar. Miriam was a fighter and he wanted to be a fighter, too. She fought for him. She fought every injustice. She was outraged about Palestine, and outraged by the immigration policies of the U.S. State Department. She encouraged Mokhtar to be vocal. To be involved. She had no fear. Any wrong, local or global, only emboldened her. It was stasis and silence that she couldn't stand, and every time he saw her, as they sat in the Infinity lobby talking about dreams, or dreams deferred, he felt stronger, more inspired and worse about his current life, opening the doors for wealthy strangers.

Especially given the existence of the button. It was right next to the phone and had been there all along. When he pushed it, the two glass doors twenty-two feet away opened. The system was quick, quiet, elegant. Using the button, Mokhtar could see a guest coming down the sidewalk and have those doors wide open and ready by the time they arrived. Even better, the button opened *both* doors. By hand, Mokhtar couldn't open both doors. They were too heavy and too big. With the button, though, the resident could stride through a fantastically wide and welcoming gateway of glass, unobstructed. They could enter the lobby, and Mokhtar, the Lobby Ambassador, could greet them. He'd be happy to greet them. It cost him nothing to look up and say hello. But to leap from the desk, to rush over, eager and panting, only to push open a door that could be opened with a button—it was a self-evident outrage and an assault on his pride. Especially when the residents passed through the lobby, entered the elevators and flew up, to apartments high above him, places he'd never seen.

BOOK II

CHAPTER X

THE STATUE

MIRIAM TEXTED ONE DAY. *You ever look across the street?* she asked. Mokhtar didn't know what she was talking about. *Across the street there's a statue of a Yemeni dude drinking a big cup of coffee,* she told him. She'd just made a delivery from her dad's deli to the building across the street from the Infinity, and in the courtyard she saw an enormous statue of a man in a thobe with a giant mug lifted to his lips. *That's got to mean something,* she said. *Maybe this is your thing.* What she meant was *You're twenty-five, Mokhtar. Pick a direction for your life.*

He'd been working a hundred and twenty feet away from this statue but he'd never seen it. It was enormous, about twenty feet tall. The man was in midstride, drinking from a giant coffee cup. Mokhtar wasn't sure about its historical accuracy—the man seemed to be some mash-up of Ethiopian and Yemeni—and why were there cute little flowers all over his thobe? It looked like he was wearing a shower curtain or muumuu. No self-respecting Arab would be wearing flowers all over his thobe.

But Mokhtar walked into the building, into the lobby across the street from his own lobby, and an encompassing history of coffee in the United States was presented in framed photos and captions. The building had been built by the Hills brothers, Austin and R.W., who in the late 1800s set up a coffee-importing business called Arabian Coffee and Spice Mills. The brothers brought beans to California from around the world and roasted them for distribution all over the West.

But freshness was a challenge. Every day on oceans or rails and roads, the coffee grew staler. This changed in 1900, when R.W. stumbled upon a way to remove the air from packaging. This became known as vacuum packing, and kept coffee beans fresher for longer, and soon revolutionized the business of coffee. The Hills brothers became phenomenally successful and were instrumental in popularizing coffee in the United States. The graphic version of the statue became their well-known logo, and the company thrived independently for one hundred years. Eventually, long after the brothers died and ceded the company to descendants and strangers, Hills Bros. was sold to Nestlé. Who sold it to Sara Lee. Who sold it to Massimo Zanetti Beverage USA. The company left San Francisco in 1997, moving its headquarters downstate to Glendale.

But the statue remained, and Mokhtar left the courtyard in a daze. Coffee and Yemen. Some ghost of a memory passed through him. That night, on Treasure Island, he mentioned the statue to his mother. She laughed.

"We've had coffee in our family for hundreds of years," she said. "Don't you remember your grandfather's house in Ibb? He had coffee trees in his yard. He's *still* got them. Don't you know Yemenis were the first to export coffee? Yemenis basically invented coffee. You didn't know this?"

* * *

Mokhtar went on a research binge. At home, on his phone, he dove in and quickly encountered a long-running debate about the origins of coffee, and the dual claims, between Ethiopia and Yemen, to its discovery.

There was widespread agreement that the earliest origin myth involved an Ethiopian shepherd named Khaldi. Apparently Khaldi was far afield with his sheep, allowing them to graze on any vegetation they could find. Every night he slept near them, and all was peaceful until late one night, when he expected them to be resting, he found that his sheep were still up and about. More than up and about—they were jumping, prancing, braying. Khaldi was mystified. He thought they might be possessed. But soon it became clear that they'd been eating beans from the bushes nearby. These were coffee beans. And when Khaldi ate the beans himself, they had the same effect on him—he was shot through with new vigor and mental acuity. He had discovered the coffee bean.

Wait. No. Not coffee beans, Mokhtar noted. The goats had been eating coffee *fruit*. Coffee beans were *inside* the coffee fruit, which grew on lush green bushes. At its most ripe, the coffee fruit was red and looked like a grape. Mokhtar saw the photos online, piles of red cherries like huge ruby-red beads. Coffee was a fruit! Mokhtar remembered this, plucking red cherries from a small tree in his grandfather's garden. The cherry could be eaten. He remembered eating the flesh of the fruit, which was sweet, and then spitting out the seeds. The seeds *were* the coffee! It all made sense now. Coffee was a fruit, from a

59

tree, a tree that usually bloomed once a year, and inside each fruit was the coffee bean. And the two halves of the bean were what we typically saw—the tiny bean, oval and with a stripe of concavity down the middle. Two halves of a bean, wrapped inside a fleshy fruit the size of a grape.

But first the bean had to be separated from the fruit. There was the red skin. Then the white flesh. Then, attached to the beans, was the mucilage, and then the silverskin. The bean inside was green, sometimes yellow, and hard like any seed. A coffee tree could be grown from any unroasted coffee bean! Of course. Did anyone know this, or remember this? If Mokhtar didn't know any of this, who did? And who knew about the Yemeni role in it all?

Few knew coffee had been born in Arabia. There were two kinds of coffee, robusta and arabica, but it was arabica that was considered far superior in taste, and it was called arabica because it was born in Arabia, specifically in what the Romans had called Arabia Felix—"Happy Arabia." This was Yemen. According to legend, it was in Mokha, a port city on the Yemeni coast, that the bean was first brewed. For centuries after Khaldi the shepherd had come and gone, Ethiopians chewed the beans and made weak tea from them, but it was Ali Ibn Omar al-Shadhili, a Sufi holy man living in Mokha, who first brewed the bean into a semblance of what we now recognize as coffee—then known as *qahwa*. He and his fellow Sufi monks used the beverage in their ceremonies celebrating God, which lasted long into the night. The coffee helped bring them to a kind of religious ecstasy, and because the Sufis were travelers, they brought coffee to all corners of North Africa and the Middle East. The Turks turned *qahwa* into *kahve*, which became, in other languages, *coffee*.

Al-Shadhili became known as the Monk of Mokha, and Mokha became the primary point of departure for all the coffee grown in Yemen and destined for faraway markets. Mokha itself was a barren and dry coastal area, not suited for coffee cultivation, but nevertheless the word *mokha* become synonymous with coffee. The coffee was grown in the interior of the country, in the mountains, using ingenious irrigation and terraces. The cherries were brought to Mokha for processing and export, and Mokha became a thriving commercial center—not just for coffee, but for other fruits and goods, too. But coffee drove trade traffic to the port, and was considered so valuable that exporting coffee plants was a crime. Men had been arrested and executed for the high treason of trying to leave port with a seedling.

The first coffeehouses, called *qahveh kaneh,* appeared throughout Arabia and were known for lively discussions, for music, and in some cases, activities frowned upon in various quarters—prostitution, gambling and criticism of local government. Coffeehouses were often closed by rulers who saw in them the beginnings of uprisings. In 1511, Khair-Bey, then the governor of Mecca, got wind that verses lampooning him had originated at coffeehouses, so he decreed that all coffeehouses be closed. But the ban didn't last long. The demand was too great.

Who knew all this? Mokhtar wondered. He could ask anyone on the street where coffee had been born, and they might say Paris. They might say Africa. They might say Colombia or Java. But who would say Yemen? What the world knew about Yemen, now, was terrorism and drones. Since the bombing of the USS *Cole* off the coast of Aden, Mokhtar had seen his parents' country devolve from Happy Arabia to what some considered one of the world's most menacing places, home to burgeoning al-Qaeda and ISIS cells and the relentless American drone strikes meant to neutralize those threats.

And the coffee trade in Yemen was all but finished. Though Ethiopia had been home to the first coffee bush, and Yemen home to the first cultivation and organized coffee trade, in the last fifty years Ethiopia had come to dominate the region. Ethiopia was now the fourth-largest producer in the world, while Yemen was all but forgotten, its exports negligible and considered of wildly unpredictable quality. In the mid-1800s Yemen exported seventy-five thousand tons of coffee a year, and by the twenty-first century produced only eleven thousand—and only about 4 percent of that was specialty coffee quality. And beyond the quality issues, Yemen was far more difficult for Western travelers. The mountainous coffee-growing regions were informally governed by local tribes and militias, whose movements were generally precarious for visitors, exporters, anyone. Given the choice between trading with the Ethiopians and the Yemenis, most coffee specialists found Ethiopia a far easier and safer bet.

The second factor was qat. Mokhtar knew qat, enjoyed qat. Qat was illegal in the United States, but in Yemen it was a central part of daily life for virtually every man. A long leaf that when chewed in significant quantities provided a mild narcotic effect, qat grew in similar climates to coffee, but was far more profitable. The incentive, then, for any Yemeni farmer to grow coffee was negligible. Most of it was exported to Saudi Arabia for a middling profit, while qat garnered higher prices and was sold locally. Given the realities of the market, coffee had been relegated to a relatively small group of passionate but ill-trained Yemeni farmers.

The training was the last and most important factor. Because it wasn't profitable for most farmers, the careful processes for the cultivation and harvesting of high-quality coffee in Yemen had been long lost. Now it was picked and stored without great care, and Yemeni coffee, though it was the world's first ever cultivated, was known to be inferior to most or all other coffees in the world.

CHAPTER XI

THE PLAN
PART I

MOKHTAR WAS GRATEFUL TO Miriam and thanked her by boring her, and Justin and Jeremy and his family on Treasure Island, with daily breathless updates about his plans to become a coffee importer. He put it all down on paper. Not regular paper. He used a large roll of white paper, the kind usually hung from an easel, and he carried his roll of white paper around every day for months, recording notes and plans and unrolling it for his friends, explaining not just the history of Yemeni coffee but his role in resurrecting it. He started with a SWOT chart—any serious endeavor in 2013 started with a SWOT chart.

Under "Strengths" he wrote:

- Highest amount of coffee genetic diversity
- Ideal microclimate
- Highest elevation
- Historical significance

Under "Weaknesses" he wrote:

- No infrastructure
- Lack of data

- Lots of defects
- No traceability

For "Opportunities" he wrote:
- Historical significance
- No one else focused on specialty coffee in Yemen
- Finding and reviving ancient varietals

For "Threats" he wrote:
- Al-Qaeda
- Corrupt government
- Pirates in Red Sea
- Tribal violence
- Andrew Nicholson (?)

Who was this Andrew Nicholson? Every time Mokhtar looked into coffee in Yemen, he ran across this name, Andrew Nicholson. Apparently he was an American from Louisiana, who had for whatever reason moved to Yemen, to the capital, Sana'a, and started to export Yemeni coffee under the name Rayyan—Arabic for "the gate of paradise." Nicholson seemed to occupy the territory Mokhtar hoped to inhabit. But having another American in the business, in Sana'a, might be enormously helpful. Economies of scale, sharing contacts, resources, camaraderie.

"This is it," he told Miriam. "I will resurrect the art of Yemeni coffee and restore it to prominence throughout the world."

Oh Jesus, Miriam thought.

But she was supportive. Everyone was supportive. Mokhtar's friend Giuliano was especially on board. Mokhtar had met Giuliano during his freshman year. Giuliano was a unicorn: a teenager who'd converted to Islam on his own. He'd been brought up in an Italian Catholic family in North Beach, by divorced parents. There wasn't much money, but the family was content, and Giuliano was a happy and curious child. His parents were greatly confused—but ultimately unsurprised—when their only son told them he was converting to Islam. He was fifteen, and had learned most of what he knew about the faith by reading *Islam for Dummies*.

His attraction to the religion started a few years earlier, when people started assuming he was an Arab. *You look Muslim,* they'd tell him. *You an Arab?* Arabic speakers would greet him with an earnest or offhand *Salaam alaikum.* Finally Giuliano looked in the mirror to get a sense of what they were seeing. *There's something there,* he thought. *Maybe I do look Middle Eastern.* That was the beginning of it, the strange catalyst for his entry to Islam: in a roundabout way, he became a Muslim because so many people assumed he was a Muslim. So he studied Islam and converted himself. Islam allows a follower to self-declare adherence to the faith, to become a Muslim by personal commitment, without any formal ceremony, so one day he declared himself a Muslim and spent his first Ramadan at Burger King.

Islam was only one point of Giuliano's connection with Mokhtar, though. As high school students, neither had much money to spend, so they found in each other a common ability to find free entertainment in the city. They would go to the Wharf to hassle tourists; they would look for dropped dollars. But mostly they talked about books and food. Giuliano, raised by Italian parents, knew food and would bring Mokhtar home for homemade risotto. They'd talk Herodotus and Edward Said and pretend they understood Plato's *Republic*. Mokhtar and Giuliano were autodidacts, and through food were awakened to heretofore unknown parts of the world and history. They'd go to Giuliano's father's restaurant—he owned one briefly, Michaelangelo's Café; it failed, and he went back to work as a waiter—and on the menu they saw sun-dried prunes, and that would be the impetus for research: Where were sun-dried prunes grown? Tuscany? Was that in France or Italy?

They taught themselves history, philosophy, and, left unsupervised for long stretches, they grew up quickly. When he was nineteen, Giuliano fell in love with a Pakistani American woman named Benish. She was brown eyed and beautiful, also a San Francisco native—they'd met right after high school—and though they knew they wanted to be married, Giuliano knew his parents, and hers, would consider it too soon. Or worse: Giuliano assumed there would be some insurmountable cross-cultural divide. Would her Pakistani father allow his daughter to marry a nineteen-year-old Italian Muslim convert? Would there be serious trouble ahead, some discussion of honor killing? (Swimming in love, his mind went to some strange places.) But Giuliano's parents readily acquiesced, and when Giuliano asked Benish's father, he gave his consent, and asked for grandchildren. Giuliano and Benish were married at her home—Mokhtar brought frankincense and myrrh—and they moved into a North Beach flat. Their first child, Saudah, was born three years later.

By then Mokhtar was working at the Infinity, and Giuliano was driving for Uber. They would unwind after work by lifting weights at 24 Hour Fitness, and before the workout, they'd drink coffee. Giuliano had grown up around coffee, and educated Mokhtar about how the Italians liked it—standing at the counter, sipping espresso, a little sugar, never milk. He took Mokhtar to the new Blue Bottle

Coffee in the Ferry Building. "This is the closest thing to a real Italian espresso," Giuliano told him, and they'd stand there, looking as Italian as they could while they downed two or three espressos to get themselves amped for heavy lifting.

That Blue Bottle was a stone's throw from the Hills Bros. building, where coffee had been imported, roasted and sent all over the western United States. The coincidences, Mokhtar felt, were adding up and making it ever more clear and irrefutable that this was destiny, that he had found his calling. No. It was more than a calling. Mokhtar referred to it as a *mission* in those early days, and he was careful not to say that he was being guided by God. But he believed this.

He pictured himself careening through the Yemeni countryside, bringing knowledge and wealth to the farmers and leaving with beautiful red cherries for export. His new life would be one of planes and horses and ships, and his story could join the pantheon of coffee explorers, those who brought about the proliferation of coffee cultivation and the popularity of the beverage around the globe. As he'd been walking around with his SWOT scroll, he imagined himself as part of the historical continuum of coffee, a vivid time line animated by a succession of rogue adventurers who also happened to be, almost without exception, thieves.

First was Bada Budan. A Muslim holy man from the Chikmagalur district of Karnataka, India, in the 1500s, Budan went to Mecca to perform the hajj. On the way back, traveling through Yemen, he encountered coffee, by then known as "the wine of Islam." Enchanted, he wanted to bring it back to India, but this was not permitted. The

Arabs would sell him as many roasted beans as he could buy and carry, but they wouldn't give him a seedling, not even a cherry.

So he stole them. He strapped seven cherries to his belly and wrapped his robe over them loosely, the folds hiding his treasure. In India, he planted the seeds in the Chandragiri Hills, and from those seven cherries, millions of arabica plants flourished. India now is the world's sixth-largest coffee producer, and Baba Budan is considered a saint.

The Dutch, too, wanted to leave Yemeni shores with a coffee plant. Coffee had first come to Europe in 1615, when it was exported from Mokha to Venice and used for medicinal purposes. Eventually it was imbibed socially and proliferated throughout parts of Europe—with the Venetians holding a monopoly on trade with Mokha. This didn't sit well with the Dutch, then a global power in naval trade. It was absurd that a commodity of this value was cultivated and controlled by so few, in one small port in Arabia. So in 1616, a Dutchman named Pieter van den Broecke, who had visited Mokha while working for the Dutch East India Company, successfully stole seedlings from Mokha and secreted them to Holland, where they were installed at the Hortus Botanicus in Amsterdam.

The seedlings took root in the garden, but the Dutch climate wasn't right for large-scale cultivation of the plant. It wasn't until 1658 that coffee was brought to the Dutch colony of Ceylon and later to Java, also a Dutch territory, where it thrived. Java soon became the primary supplier of coffee to Europe, and Mokha's primacy waned.

The Dutch were as careful with their monopoly as the Yemenis had been, assiduously protecting the farms in Java, blocking any export of seedlings or cherries. For half a century the Dutch enjoyed control of the European market, until the French entered the business via a bizarre act of economic self-harm on the part of the mayor of Amsterdam. In 1713, he presented King Louis XIV with a coffee plant. It was to be a gift, he insisted, not the beginning of an industry, and for years the French observed this understanding, keeping the plant within the walls of the Jardin des Plantes in Paris. Visitors could admire the plant from a distance, and most did so without any thought of subterfuge or theft. A man named Gabriel de Clieu, though, had different plans.

De Clieu was an officer in the French navy and was determined to bring coffee to the West Indies, a French territory that was considered a suitable coffee-growing counterpart to Java. He set sail in 1723 on a corvette called the *Dromadaire,* but two weeks into the trip, his ship was attacked by pirates off the coast of Tunisia. The *Dromadaire* was well armed, though, and with its twenty-four guns was able to repel the pirates. They were only a few hundred miles from Martinique when the ship was damaged by a storm and began to take on water. Cargo had to be jettisoned to prevent the *Dromadaire* from sinking, and among the jettisoned cargo was a good deal of the crew's drinking water. For the rest of the trip, water was strictly rationed, and de Clieu had to share his own small ration of water with the coffee plant, drop by drop. On Martinique, de Clieu planted his seedling, which begat hundreds more, which he distributed across the island. From there coffee cultivation grew almost exponentially, replacing the island's previous cash crop, cocoa. De Clieu was a hero, and the French had a monopoly on coffee cultivation in the Western Hemisphere. For a time, at least.

Francisco de Melo Palheta was a lieutenant colonel in the Brazilian army, Brazil at the time still under Portuguese control. The Portuguese badly wanted in on the rapidly expanding market for coffee, and they saw Brazil as a perfect environment for growing the plants. But they had been unsuccessful in getting their hands on a seedling.

By this time, the French were cultivating coffee not only in Martinique but in French Guiana, too, and in 1727, that colony became embroiled in a border dispute with Dutch Guiana, the territory just over the Rio Oiapoque. To settle the matter, the two colonies asked the ostensibly impartial Brazilians to intercede, and Brazil sent Francisco de Melo Palheta. By then Palheta was fifty-seven, but he was still a handsome and romantic man whose charms had certain effects on the women he encountered. He traveled to Cayenne, the capital of French Guiana, where he sat with the French and Dutch colonial governors and settled the question of the border. But that was not his primary goal. During his time in Cayenne, he was conspiring to get a seedling out of the country. The farms where the coffee plants were grown, though, were well guarded, and he was a known figure, so he couldn't be seen skulking around.

Instead, he went about seducing the governor's wife, Marie-Claude de Vicq de Pontgibaud. She was so taken with him that at a state dinner held in his honor, to thank him for brokering an agreement on the border, she provided him with a bouquet of flowers, inside of which she'd hidden enough coffee cherries to start a farm of his own.

He planted the first coffee plants in the Pará region of Brazil, and within seven years he had a thousand bushes thriving. These plants became the foundation of the Brazilian coffee industry, which by 1840 accounted for 40 percent of the world's production. One of Brazil's largest markets was the burgeoning colonies of North America. The Dutch had introduced coffee there in the 1600s, and it was reasonably popular, always sharing primacy with tea. But as tensions grew between the colonists and the British Crown, and as taxes on tea grew ever more onerous, the colonists began to see tea as emblematic of the British yoke.

On December 16, 1773, hundreds of colonists, most of them dressed as Native Americans, met four ships of the British East India Company in Boston Harbor, and they dumped all the tea on board the ships into the sea. Tea drinking in the United States was never the same. Coffee became the stimulant of choice for the new nation, bought primarily from the Dutch—thus the moniker "java." Its popularity grew in spurts until the twentieth century, when mass production, better storage and packing techniques—Hills Bros. being instrumental in much of this—and demand wrought by World War I and World War II all conspired to make the United States the world's leading coffee consumer. By the twenty-first century, Americans were consuming 25 percent of the world's coffee, and by 2014 coffee was one of the most valuable agricultural products in the world, a seventy-billion-dollar business, with the cherries grown in Colombia, Vietnam, Cambodia, Kenya, Uganda, Guatemala, Mexico, Hawaii, Jamaica and Ethiopia.

But Yemen, the region that had started the cultivation of the coffee plant in the first place, was now a tiny and largely ignored player in the world coffee market. Mokhtar had the idea that he could change that. But first he had to see Ghassan Toukan.

CHAPTER XII

SAGE ADVICE FROM GHASSAN TOUKAN

PART II

GHASSAN TOUKAN LOOMED LARGE in Mokhtar's mind. With any talk of money, or starting a business, he first thought of Ghassan.

After his stint tutoring, or trying to tutor, the young Mokhtar Alkhanshali, Ghassan went to San Jose State University thinking he'd learn everything he needed to know to launch his own tech start-up. But the pace was slow and the professors at the time, most of them converted math professors, were out of step. They couldn't teach him what he wanted to know. Ghassan dropped out and started his own consulting business, building and improving computers for friends. Meanwhile, he worked at a cell-phone shop in San Francisco, on Market Street. His parents expected their son to get an undergraduate degree, and perhaps a master's. They were aghast to see their dropout son working at a cell-phone store on a shady block of mid-Market.

But Ghassan had designs. He and a friend had built an e-commerce platform and created their own company, which was acquired by an e-commerce behemoth for a respectable sum. Ghassan was not yet a landed man of leisure, but he had done extraordinarily well, and Mokhtar had been paying attention. He and Ghassan had stayed in touch over the years, and now that Mokhtar had this coffee notion, he wanted to talk it over with the most successful entrepreneur he knew.

They agreed to meet at Four Barrel Coffee in the Mission. Ghassan arrived first, expecting Mokhtar to be late, but Mokhtar was uncharacteristically on time. And he was carrying some kind of picture frame. *Was he really carrying a picture frame?* Ghassan wondered. He was. Mokhtar had brought a framed picture to the meeting. It was enormous.

"Check this out," Mokhtar said, and unveiled it. It was a reproduction of an English-language newspaper from 1836. On the front page there was an engraving of the ancient port of Mocha. Mokhtar launched into a long and meandering monologue about coffee, Yemen, the port of Mocha, the two spellings—which spelling did Ghassan like, by the way, Mocha or Mokha?—how he'd discovered his connection to all this, and how he planned to become a coffee importer-exporter and revive the ancient art and prominence of coffee from his ancestral land. Ghassan didn't know what to say. Mokhtar was all over the map.

"Do you have a business plan?" Ghassan asked.

Mokhtar unveiled his business plan with the same flourish he'd given the framed newspaper. It was a stack of multicolored paper an inch thick—a bizarre combination of manifesto, history lesson, idea dump and rant.

"And this," Mokhtar said, pointing to a page full of bullet points. The bullet points alone, Mokhtar implied, made it a business plan, and made it a great one.

Ghassan looked at it. He tried to read it. Finally he took a breath and said, "Mokhtar, I have to be honest with you. This is the ghettoest business plan I've ever seen."

Still, he knew Mokhtar was onto something. He saw the passion in his eyes, on the pages. The business plan had to be rebuilt from the ground up, but there might be something there. They would have to change the name. The Monk of Mocha was wrong with either spelling. Who was the monk? Was Mokhtar the monk? Why was Mokhtar suddenly a monk?

"No. It's not me," Mokhtar said. "It's this guy from this book I have . . . Hundreds of years ago, there was a monk in the port of Mokha who—"

"Forget it. Forget the monks," Ghassan said. "No monks. Focus on coffee. Focus on the business. Come to think of it, you have to make a choice. Are you a businessman or are you an activist? For now, at least, you have to pick one."

Mokhtar's pages were full of educational aspirations, dreamy language about cross-cultural collaboration, teaching the world about the beauty of Yemen, Yemen beyond terrorism and drones.

"But this isn't a nonprofit," Ghassan said. "Start a real business, and all that will happen. Education about Yemen will come through customers' engagement with the product. And in the meantime you'll employ actual Yemeni people. And you'll do something tangible. And you'll make a living. And you won't have to ask for donations. And it won't have to be about Islam. You're not selling Islamic coffee beans. Sell Yemeni beans. Do that, and do it well, and the rest will follow."

Ghassan left the meeting, and a few days later went to Mecca for pilgrimage, and from there to Japan—it was cherry blossom season and he loved the cherry blossoms—and all the while he thought about Mokhtar and his business plan.

Ghassan knew about coffee. He'd gotten sucked into the specialty coffee world years before; it was near impossible in San Francisco to avoid it, difficult to avoid becoming at least a dilettante, just as it was impossible not to become passably knowledgeable about technology or wine. But with coffee, Ghassan was only a customer; he'd never had any interest in the business side of it. In fact, he'd spent years convincing a dozen different friends *not* to open coffee shops. Mokhtar wasn't the only one who'd come to him for business advice, and over the years an alarming number of those who sought his help were planning to start cafés.

No, Ghassan had said to every one of them.

They wanted to create community spaces, spark the next Enlightenment, bring people together in an atmosphere of—

No. No, no, no, he said. *No.*

This was how he spent his time: convincing otherwise sane and successful former techies not to start cafés. There was almost no way to make one profitable, he told them. And a café in San Francisco? High rents and low margins. Your customers will be problematic. Some guy with strenuous facial hair sitting for six hours at one of your tables, fondling his laptop and drinking one cup of coffee, for which the margin was, what, twenty cents? It couldn't work. The only way to make money in coffee, he told all these would-be café owners, was to buy the green beans, roast them and sell them—control the supply chain, set prices, get the beans from the point of origin. That's where the margins are.

But no one ever wanted to do that.

No one until Mokhtar. So while flying in and out of Saudi Arabia, and while walking under the cherry blossoms in Kyoto, Ghassan thought Mokhtar might really be onto something. He knew that Yemeni coffee was supposed to be good, but that it was difficult to get out of the country. If a Yemeni American went to Yemen, wouldn't he be a kind of natural bridge between the inaccessible mountains and political mess of Yemen and the world market for these beans?

After so much time in the ephemeral world of software, Ghassan was looking for something more three-dimensional. Coffee could be smelled and tasted and touched. And it was a commodity, recession-proof. Next to gasoline, it might be one of the most recession-proof commodities of all. Fuel for the machines, fuel for the people.

"But get serious," he told Mokhtar that day in the Mission. "At least know what the hell you're talking about."

CHAPTER XIII

PAST PRETENSE

MOKHTAR KNEW ABOUT BLUE Bottle. Giuliano had brought him
to Blue Bottle. He'd been hearing the name now for years, and Blue
Bottles were emerging all over San Francisco. Ever since he started
talking about his future in coffee, people had been telling him to go
to Blue Bottle, study at Blue Bottle, and he planned to, but first,
because he was a man of research and erudition, he dug deeper and
found another story of adventure, another man risking his life to get
coffee from one place to another.

In 1683, the Ottoman Empire was at the height of its power,
occupying a huge swath of eastern and central Europe. The Ottoman
Turks, wanting to overtake Vienna, surrounded the city with three
hundred thousand troops. The city had little hope of withstanding
the Ottoman attack unless the Viennese could send an envoy through
the enemy lines to get help from the Polish army 287 miles away. The
Polish army could attack from the rear, the Viennese from the front.

The Viennese elected from among their ranks a young Pole named
Franz George Kolshitsky, who had spent time in the Arab world and
spoke Arabic and Turkish. The Viennese dressed him in a Turkish

soldier's uniform and sent him through the night, across enemy lines. He made it to the Polish troops and delivered the message. The Poles came to the aid of Vienna, and together they drove back the Ottoman siege. In their retreat, the Turks left much of what they'd brought with them, including twenty-five thousand tents, five thousand camels, ten thousand oxen and five hundred bags of small, hard green beans.

The Poles assumed the beans were camel feed, but Kolshitsky knew better. These were coffee beans; in the Arab world he'd seen them roasted and brewed. As a reward for his heroism, he was allowed to keep the beans, and with them he opened the first coffeehouse in central Europe, calling it The Blue Bottle. There he made coffee as he'd learned in Istanbul, and awaited success. Success did not come. The Viennese didn't take to this new beverage. It was too strong, too bitter. To blunt its edge and save his business, Kolshitsky added a spoonful of cream and a bit of honey. Now the crowds arrived. His concoction was replicated and disseminated. He'd invented Viennese coffee and had brought coffeehouses to Europe.

About 320 years later, there appeared an American named James Freeman, who among coffee eccentrics seemed uniquely qualified to be king. He had once been a professional clarinetist—second chair in the Modesto, California, symphony. He was also a home-brewing coffee hobbyist, a purist frustrated by the ever-permutating bastardizations of coffee—the pumpkin-spiced lattes, the caramel macchiatos. He wanted to return to the basics, to allow customers to taste the actual coffee, brewed in front of them, cup by cup. He harbored dreams of building a larger roaster, which would combine elements of an adobe oven and a rotating drum—and would be powered by either a human (or a dog, he noted) running on an attached treadmill. He took his design to various planning and health officials in Oakland, who were bewildered and unamused.

Eventually Freeman settled into roasting on a Diedrich IR-7, manufactured in Sandpoint, Idaho, and powered by standard electricity. He set up shop in San Francisco's Hayes Valley, where he pioneered a very slow and methodical way to make coffee, every cup its own unique undertaking, drip by drip. His little shop quickly evolved from neighborhood curiosity to something with a cult following. He called it Blue Bottle.

Blue Bottle's headquarters were now in Oakland's Jack London Square. Every Sunday, Blue Bottle ran an open cupping session, where anyone could come and witness and participate in cuppings, analyzing the tastes of various brewed coffees.

Ghassan couldn't come that first time, so Mokhtar brought Omar Ghazali, to whom he still owed three thousand dollars and who, he hoped, might see an opportunity. Omar knew fruit, and coffee was a fruit, and Omar knew start-ups, and knew Yemen. With the cash from his fruit business, Omar had invested in a T-shirt business, a sheep operation, a phone-card scheme. He was open to new opportunities.

When they arrived that Sunday, there were about a dozen other people assembled, and though Mokhtar feared, and expected, that it would be a highly pretentious place, he and Omar found it welcoming and largely free of attitude. The Blue Bottle staff had set out about forty cups on a high table, each filled with a different coffee, each a different roast and varietal. Then they demonstrated how they assessed the taste and quality of every coffee by taking a spoon to each cup, bringing it to their lips, and then, instead of just drinking that spoonful, they slurped it. It had something to do with oxygenating the coffee, bringing out its full flavor. Thus they would slurp from each cup, swish it a bit in their mouths and then, using another, taller cup that they carried with them, they would spit out each spoonful.

Mokhtar said nothing. He just watched. But he could tell Omar wanted to smile or laugh or walk out and never come back. The man in charge walked from cup to cup, slurping—loudly—and then spitting, and it was impossible to imagine how this could possibly lead to a better assessment of any coffee. Why not drink it? Why not drink more than a spoonful? And wasn't the slurping distracting on some elemental level?

But then it was his turn, and Mokhtar took his spoon to the cup in front of him. He let a small pool of the brown liquid fill the spoon, and he brought it to his lips, wondering as he did what kind of sound his slurping would make when he actually slurped. When he did, the sound was quick and high-pitched, and though he expected that at least someone in the room would be laughing when he looked up, no one was, and he swished the coffee in his mouth and tried to think what it tasted like.

Was it toasty? He wrote down *toasty*. Was it fruity? He'd heard the word *fruity* a lot that day, so he wrote down *fruity*. Nearby, someone said they tasted chocolate notes, and Mokhtar said he did, too. The class veered between the practical and the impenetrable. It lasted an hour and included far too much information to absorb—there was talk of varietals, and flavor notes, and first crack, second crack, of light roast and dark roast and Guatemalans and the five layers of the coffee fruit.

Mokhtar's head felt heavy and his soul was discouraged. He was masterful at taking in vast amounts of information and regurgitating it quickly, but this was too much. Still, afterward he felt compelled to approach the man who had led the class, Thomas Hunt, to tell him his plans. He told him he was from a Yemeni family that had been growing coffee for centuries and that he would soon return to Yemen to revive the art of Yemeni coffee and bring it to the specialty coffee market. Thomas, while being moderately encouraging, mentioned that Yemeni coffee had a reputation for being dirty and inconsistent and that getting the coffee out of the country had challenged any number of experienced exporters before Mokhtar.

I can make it better, Mokhtar thought, *and I can get it out.*

He had no reason to believe either was true or possible.

Mokhtar returned to Blue Bottle the next week, and this time he brought Justin, who was still considering a foray into olive oil, and the two of them took notes, and cupped, and learned a little more, and again, when the class was over, Mokhtar hung around and reintroduced himself to Thomas and reiterated that he was serious about reviving Yemeni coffee and bringing it to the world's specialty coffee stage, and creating cross-national cooperation, introducing a different idea of Yemen to the world, a Yemen apart from drones and al-Qaeda. And this time Thomas, whether believing in Mokhtar or wanting to get rid of him by passing him on to someone else, mentioned a man named Graciano Cruz, a Panamanian who was doing the same kind of thing but for coffee in Ethiopia, Peru and El Salvador.

"You should talk to him," Thomas said.

"How?" Mokhtar said, feeling sure that this name, Graciano Cruz, was the next keeper of secrets along his hero's journey.

"I'll send you his e-mail address," Thomas said.

But Thomas didn't send him his e-mail address.

Every week, Mokhtar went back to Blue Bottle, cupped and learned and lingered, and copied everything written on any white-board, and every time, after the session, he asked Thomas for the e-mail address for Graciano Cruz, and every time Thomas said, *Sorry, I spaced,* and said he would send it the next day—because he insisted Mokhtar and Graciano really should talk, their missions were aligned, they really should know each other—but every week Thomas would forget again.

Mokhtar continued to come to Blue Bottle, now on weekdays, too, and Thomas and the rest of the Blue Bottle staff allowed it and even put him to work for the public cupping, and soon enough Mokhtar had some mastery of the basics.

CHAPTER XIV

THE BASICS

THERE WAS THE COFFEE plant. He knew the coffee plant.

Coffea arabica. It was something between a bush and a tree and it seemed acceptable to call it either. Some called it a shrub. It could grow up to forty feet tall, but that wasn't ideal—ideal was smaller, six to ten feet. It needed a fair amount of water and could thrive in full sun or partial sun, and in most climates flowered twice a year, delicate white petals like certain orchids. And when it flowered, it produced cherries that went from yellow to green to red and which, when picked at just the right moment, provided the best coffee. But the beans were deep within the cherry. The cherry, oblong and bright and smooth like a grape, had five layers within it. There was the skin, the red outer covering. Below that was the pulp, an edible and even juicy layer, tougher and leaner than a grape but otherwise not so different in consistency. Below that, there was a very thin layer called the mucilage, and under that, the parchment. Below that, one more very thin layer called the silverskin and, finally, under all that, was the bean, which was really a two-headed seed varying in color from green to khaki.

The average coffee tree produced about ten pounds of cherries during any given harvest, and in most countries, every one of the cherries had to be plucked by hand and dropped into a basket carried by a picker. This was just the beginning of the process, one of the most complicated processes for any crop—quite possibly the most complex journey from farm to consumption of any foodstuff known to humankind.

The trees, first of all, needed the same kind of care and guidance as any large plant—they needed to be fertilized, protected from pests and pruned so the lower boughs could produce most of the fruit (for pickers shouldn't have to use ladders, and besides, the higher boughs produce less fruit).

Each healthy tree produced hundreds, even thousands, of cherries, and these will ripen twice a year, but not at the same time. That is, on any given bough there will be cherries of a maddeningly wide range of ripeness, and the best (and some would say only) cherries are red, picked at a peak of readiness—the redder they were, the higher the sugar content and the better the taste. Thus the pickers had to be judicious. They must pick the red cherries, let the yellow and green ripen and remove the already overripe. A good picker might be able to fill a thirty-pound bucket in an hour and about twelve buckets a day, meaning about 360 pounds of cherries a day. Ideally those buckets were all red—thousands of red cherries. That would be a day's work, each requiring an eye, two fingers, a twist.

These cherries were brought to a central depository on the farm to be processed. A small farm—there are tens of thousands of small farms around the world, many of them only a few acres—would often send its cherries to a shared mill for processing, while the larger farms would do their own processing, but in any case the point was to separate all those layers from the bean itself. That was what the word *processing* meant—the removal of all five layers from the beans. And to do that, there were two primary methods, wet and dry.

Wet processing was the most commonly used method, producing what was often called washed coffee. In this process, the red cherries were sent through a pulping machine to remove the outer skin and the pulp. This left the pulp, a viscous and slippery layer, still on the beans. The beans were soaked in water and then allowed to ferment for anywhere from hours to days. The mucilage dried and became easier to detach from the beans. This involved more water. The beans were washed again until all that was left was the green beans. These green beans were then dried for four to eight days, either outside in the sun and air or in mechanical dryers. Wet processing produced a certain uniformity of quality in the beans, which was desirable in specialty coffee, but it used an astonishing and perhaps unsustainable amount of water.

Natural, or dry processing, was the more ancient process, believed to have originated in Yemen and still practiced there. As its name promises, it involves no water. The cherries would be dried on flat beds, usually some kind of latticework like a metal screen, and when dry, they were hulled—run through a rudimentary machine to remove all the layers from the bean. Because the beans weren't washed, they retained some of the mucilage, and because the beans spent more time inside the fruit absorbing its flavors, dry processing resulted in a fruitier but far less predictable taste. For generations this has been both the bane and boon of Yemeni coffee—it could be either wonderfully rich or of such rough quality that its merits were obscured.

After the beans were processed, they were bagged and left to rest. They needed to rest, Mokhtar learned, because the processing was traumatic for the fruit—traumatic for the fruit!—and after such a traumatic process, they needed time to recover. They were still alive, remember, Mokhtar was told. They were seeds, remember. They could still produce a coffee plant. So the resting period can last anywhere between three and six months. Farmers who weren't so quality conscious would store their beans for far longer and still produce decent enough coffee, but most experts agree that storage shouldn't last more than a year—beans should be roasted within a year of harvest.

But first they needed to be sorted.

In almost all cases, in all farms or processing plants, there are rows and rows of human beings, usually women, who hand-sort the beans. Their task is simple but labor intensive: they take piles of beans, hundreds of individual beans, and painstakingly remove any that are defective. A bad bean, experts say, can be just like a rotten apple—it could spoil the entire batch.

What constitutes a defective bean? Often the defects were obvious. Some beans were broken. These fragments needed to be removed. Some beans were rotten. Or they've fermented. Or they were sour. Usually the flawed ones were obvious. The sorters, sitting at tables or desks, took pile after pile of the beans, and removed the offending beans. This process took days, and involved a level of concentration and care that would surprise the billions of coffee drinkers who assume that beans are beans—that they were all thrown together, all roasted together. But with a startling attention to detail and a commitment to uniformity, humans actually picked and chose each bean.

Then there was shipping. The sorted green beans were labeled and packed and shipped. At Blue Bottle, Mokhtar learned the packing and shipping of the specialty coffee variety, where each harvest was carefully recorded and exported in smaller quantities—measured in kilos, not tons. In specialty coffee, the farm was known. The growers were known. Bags featured the name of the country, the region—Antigua, Guatemala, for example. Then the variety: for example, Bourbon, Typica. Often they named the farm and the farmer—a level of intimacy and information akin to wine or fine cheese.

These bags, then, were sent to roasters. Blue Bottle was a roaster. Royal Grounds was a roaster. Intelligentsia was a roaster. These were the individuals or companies—roasters could be large or small, multinationals or microscopic hobbyists—that took the raw green beans and heated them until they resembled the beans we associate with coffee.

For the last century or so, what most people knew about roasting was that it was done to coffee, and the French did it one way while the Italians did it some other way. At Blue Bottle, though, Mokhtar could see it actually being done in front of him—in gigantic machines from Germany that belched heat and needed constant tending. The roasters would drop the beans from a chute into a large drumlike oven, where they were continuously rotated and churned to ensure that they were baked equally on all sides. But were all beans roasted the same way? Not at all. All beans were different. But in all cases, it was considered anathema to overroast the beans.

A good coffee should be roasted gently, in small batches, and lightly. A dark roast hid the greatness of a coffee, or steamrolled it, much like burning a steak would ruin a good cut of meat. Roasted coffee had more than eight hundred different aroma and taste components, and bringing out any respectable amount of these required the skill of an artisan. Mokhtar watched the Blue Bottle roasters do this, and the process reminded him of watching a great chef or glassblower—work requiring artistry and precision, but also the manipulation of fire, of valves and levers. And it all lasted a few minutes. The average roast only took ten minutes, and every second was crucial. Periodically, in the middle of a roast, the roaster would remove a few beans, checking the color, the size, the cracks. It was an intense few minutes, and as often as not, the roaster, even after all that, would think he or she could have done it better. Ideally, the roasted beans then were allowed to rest. They reach their flavor peak three days after roasting, and after seven, they begin to decline. Grinding the coffee three days after roasting is ideal, and it's best to brew it immediately after grinding.

Everywhere along the line there were people involved. Farmers who planted and monitored and cared for and pruned and fertilized their trees. Pickers who walked among the rows of plants, in the mountains' thin air, taking the cherries, only the red cherries, placing them one by one in their buckets and baskets. Workers who processed the cherries, most of that work done by hand, too, fingers removing the sticky mucilage from each bean. There were the humans who dried the beans. Who turned them on the drying beds to make sure they dried evenly. Then those who sorted the dried beans, the good beans from the bad. Then the humans who bagged these sorted beans. Bagged them in bags that kept them fresh, bags that retained the flavor without adding unwanted tastes and aromas. The humans who tossed the bagged beans on trucks. The humans who took the bags off the trucks and put them into containers and onto ships. The humans who took the beans from the ships and put them on different trucks. The humans who took the bags from the trucks and brought them into the roasteries in Tokyo and Chicago and Trieste. The humans who roasted each batch. The humans who packed smaller batches into smaller bags for purchase by those who might want to grind and brew at home. Or the humans who did the grinding at the coffee shop and then painstakingly brewed and poured the coffee or espresso or cappuccino.

Any given cup of coffee, then, might have been touched by twenty hands, from farm to cup, yet these cups only cost two or three dollars. Even a four-dollar cup was miraculous, given how many people were involved, and how much individual human attention and expertise was lavished on the beans dissolved in that four-dollar cup. So much human attention and expertise, in fact, that even at four dollars a cup, chances were some person—or many people, or hundreds of people—along the line were being taken, underpaid, exploited.

CHAPTER XV

THE C MARKET AND
THE THREE WAVES

THE PROBLEM, MOKHTAR REALIZED, was the C market. Coffee was a commodity, and the price paid for the vast majority of the coffee harvested and sold worldwide was dictated by what the C market said coffee was worth. If the C market announced that coffee was going for one dollar a pound, then that price drove what farmers anywhere—from Guatemala to Rwanda to Vietnam—could charge for their crop. Of course, the farmer himself isn't receiving that one dollar. That's the final price paid by the conglomerates—Nestlé, Procter & Gamble, Philip Morris and Sara Lee—that buy 40 percent of the world's coffee. The average small farmer in, say, Colombia might actually sell his coffee for thirty cents a pound. And not to the conglomerates directly. No, first he might sell his crop to a loan shark—a local broker who lends the farmers money until harvest, essentially keeping the farmers in perpetual debt. The loan sharks mix and bundle the crops of dozens of small farmers, and sell their aggregate to regional brokers. These brokers operate throughout Colombia, buying the crops from dozens to hundreds of small farms, consolidating them all into one undifferentiated mass called, for example, Colombian Coffee.

The brokers will then sell this hulking regional bounty to an international conglomerate for the commodity price.

This system was initiated in coffee's first wave, a period in which Hills Bros. was instrumental. During the first wave, coffee exploded in popularity, became a multibillion-dollar business subject to all the benefits and liabilities of mass production. Vacuum packing made it easier to keep coffee fresh and to bring it to far-flung places but further distancing the customer from the roaster. Satori Kato, a Japanese American, patented instant coffee in 1903, allowing Nestlé, Maxwell House and Folgers to market coffee more as a caffeine-delivery product than a foodstuff of any sensory quality. Mass-produced coffee was cheap but it tasted wretched, with consumers forced to add sugar and milk and countless other after-market codicils to make it tolerable.

Coffee's second wave was a reaction to the downward spiral of coffee prices and quality. In the 1960s, Alfred Peet opened a small coffee roastery and coffee shop in Berkeley, California, where he refocused attention on where the beans had come from and how they were best roasted. A cup of coffee at Peet's was more expensive than at the diner down the street, but it was far superior. Customers caught on, made Peet's a success, and enabled other entrepreneurs, including Howard Schultz of Starbucks, to expand the reach of coffee's second wave. Like Peet, Schultz was a socially conscious businessman, and made an effort not just to highlight where his coffee had come from, but to pay these farmers better, too. As Starbucks grew to a global phenomenon and emphasized the social space of a café—sometimes over the coffee itself—many in the coffee world wanted to return yet again to a grassroots, artisanal approach to roasting and brewing, where the emphasis would be squarely on the coffee itself.

The third wave of coffee began. Third-wave roasters were usually independently owned and operated, not chains. They highlighted the origins of their coffee not only by country or region, but by the actual farms where the coffee came from. The names of the farmers. The soil, altitude and shade factoring into the taste of that coffee. They roasted their beans on the premises and brewed those beans promptly. They preferred the pour-over method, one cup at a time, giving each cup a handmade, one-of-a-kind specificity not unlike a customer would get while sipping a new Cabernet at its winery of origin.

And the comparison to wine was, Mokhtar knew, the key to making the third wave work. When a contemporary patron at a restaurant wanted wine, they asked for a wine list. On the list were not just types of wines—Cabernet, Pinot Noir, Chardonnay—but dozens of choices for each. A discerning customer might want not just a Malbec, not just a Malbec from Argentina, but a Malbec from Vines of Mendoza winery in the Uco Valley, where the soil and water and higher elevation were considered ideal for the cultivation of this velvety red. Because wine had long benefited from this level of specificity and customer erudition, winemakers had more control over what they could charge. If they made a high-quality wine, they could charge a premium for it. Thus winemaking became something of a meritocracy, something coffee wasn't, given its shackling, since 1882, to commodity pricing.

The third wave offered to free farmers from the C market. There might be a farmer in Ethiopia who had, for twenty years, been subject to dollar-a-pound commodity pricing—a rate that kept him and his employees in poverty. But if that farmer managed to create an exceptional coffee, he or she might enter it into a regional or global competition, and if that coffee was highly rated, he or she could get the attention of a third-wave roaster, like Chicago's Intelligentsia Coffee & Tea or Stumptown Coffee Roasters in Portland. And then something extraordinary could happen. They could trade directly.

Just as the third wave built on the crucial work of the second wave, so did direct trade advance from the important work of those advocating for fair trade. While fair-trade advocates had made great strides in ensuring that the products consumed by the first world were not made by exploited humans in the developing world, direct trade took it a step further. When that Ethiopian coffee farmer trades directly with the roasters from Intelligentsia, all the pitfalls of commodity pricing fall away. All the unknowns were eliminated. The roaster might travel to that Ethiopian farm, meet its owners and staff and pickers, inspect the trees and processing plant and see with his or her own eyes what they're buying. If the quality was high, and the business practices sound, then the roaster and producer could agree on a price with no interference from loan sharks, brokers, international conglomerates or the C market. The direct-trade roaster invariably will pay more than the Ethiopian farmer has sold his beans for in the past. Freed from the merciless will of the global market, the farmer might sell his coffee for three dollars a pound, ten dollars a pound, twenty. There are rare varietals from all over the world—from El Salvador, Hawaii, Panama—that sell for forty dollars a pound. The effect was immediate and profound. If, through direct trade, the farmer gets a dollar more per pound, the transaction has radically changed the farmer's life and the lives of his pickers and staff. If the farmer gets forty times the commodity rate, then what had been a break-even endeavor becomes a profession—and everyone involved could live with dignity and pride.

The last step was convincing the customer to pay for it. A customer accustomed to a two-dollar cup of coffee will startle at the idea of paying five dollars for a cup of direct-trade Ethiopian coffee. But if the customer knows that five dollars is the actual price that cup of coffee should be—the correct price to ensure that everyone involved in bringing that cup of coffee into existence is being treated humanely and given a chance to live with dignity—would that customer balk or step up?

Mokhtar thought about the difference this could make in Yemen. In Yemen, he knew, coffee was so labor intensive, and was sold so cheap—mostly to brokers and loan sharks bringing it to Saudi Arabia by land—that it was nearly unworkable for any Yemeni farmer. Years ago, most farmers had forgone their coffee for qat. Though it needed more water, qat was far more profitable, and most of it was consumed within Yemen. There were fewer people involved in the trade, and the business was simple. Make qat in Yemen, sell qat in Yemen.

The only way Mokhtar could revive coffee in Yemen, then, was to raise the price paid for Yemeni coffee above that paid for qat. To do that, he had to deal directly with the farmers, and determine a price based on what he could get from international specialty roasters. And to garner a higher price from these specialty roasters, he had to drastically raise the quality of Yemeni coffee cultivation. And he had to begin without having set foot on a Yemeni coffee farm.

CHAPTER XVI

THE PLAN
PART II

NOW MOKHTAR HAD A new plan. A plan far better and more focused than the ghetto version he'd presented to Ghassan. He wrote this new version up, with the heading MONK OF MOCHA. He still hadn't changed the name and hadn't decided how to spell *Mocha*. But conjuring a new name was on his list. He had a lot to do.

"Vision: To empower Yemeni coffee farmers with the knowledge and tools to bring positive changes in the quality of their coffee and life."

He'd seen some business plans, and they usually started with that—the Vision summarized in one sentence. Anything more than one sentence was deemed unfocused. After Vision, the Mission allowed more elaboration. He wrote:

"Mission: To create an economically viable and sustainable coffee company in Yemen with the purpose of improving the quality, consistency and production of coffee beans which will be the vehicle for changing the lives of the growers and producers through high ethical standards and socially conscious business practices.

"Core Values:
- "Putting the farmer first
- Honesty and transparency
- High ethical standards in all affairs
- Responsibility and accountability
- Quality over quantity"

After the Mission there was usually something like Strategic Areas of Focus, so Mokhtar elaborated on that.

"Strategic Areas of Focus: Our main area of focus is the specialty coffee market. We want our growers to produce high-quality, high-consistency sun-dried arabica coffee beans with clear traceability. Our farmers will use more effective methods of growing, harvesting and processing without losing their traditional and ancient heritage of cultivating, but rather finding a place where the best of the new and old worlds meet."

He showed the plan to Ghassan.

"Better," Ghassan said.

Mokhtar had finally tracked down Graciano Cruz, and they'd struck up an online friendship. Graciano told Mokhtar about an event coming up in Los Angeles, where specialty coffee roasters and traders from around the world would gather. "I know the people running the conference," Graciano told Mokhtar. "Just tell them you know me."

Mokhtar didn't want to go alone and thought he might look more professional if he had an accomplice. He called Giuliano, but Giuliano didn't want to drive down to LA. Justin didn't want to drive down to LA, either. No one wanted to make the drive, and Mokhtar didn't have the funds for a flight, so he called up Rafik, his uncle on his mother's side. Rafik had been a cop in Oakland but now lived in Richgrove with Sitr, Taj and Rakan.

Rafik was a man of constant personal reinvention. He was just six years older than Mokhtar but had already lived a dozen lives. He'd been a security guard at the Museum of the African Diaspora. Then a UPS driver. Then an AC Transit bus driver. He'd even lived with Mokhtar's family on Treasure Island for a year. Finally he joined the police academy, excelled, won awards for marksmanship and was named his class's valedictorian. He served in Oakland for six years as a beat cop, but eventually hurt his back and was on disability. Back in Richgrove, he was considering his options. Maybe he'd open a hamburger place, or start a grape farm. Maybe a coffee shop.

Mokhtar asked him if he wanted to go to a conference about specialty coffee. Rafik, who thought of himself as a foodie, agreed. So Mokhtar, aspiring coffee importer/exporter, drove four hours to Richgrove, picked up Rafik, and the two of them drove the next few hours to Los Angeles, Mokhtar filling Rafik's head with the potential glories of Yemeni coffee, how it could save the country's trade sector and announce to the world that Yemen had more to offer than drone strikes and qat. But neither man had any idea what to expect at the conference or if they were dressed right, or if they would be asked for credentials, or any proof they belonged there. They didn't even have business cards.

The guy at the door, a young bearded man with a wide smile, asked them their affiliation. Mokhtar said he was with Monk of Mocha, a Yemeni American cross-national operation. (He still hadn't changed the name or decided on the spelling.)

"Okay," the bearded man said.

"We're resurrecting Yemeni coffee," Mokhtar said, and continued for a few minutes, talking too much, especially given they hadn't entered the building yet.

It was only once they entered the conference that Mokhtar realized that he did not belong. The three largest Ethiopian coffee exporters had come more than nine thousand miles to meet the largest buyers in the American specialty coffee market. People were there from Stumptown, Intelligentsia and Blue Bottle. Mokhtar was neither an Ethiopian coffee grower or an American coffee buyer, and any hopes of blending in, or hiding among hundreds of attendees, were quickly gone. There were only twenty people at the conference.

Mokhtar and Rafik attended panels and cuppings, pretending to belong. But Mokhtar did not feel he belonged, even after his months at Blue Bottle, and even after spending two hours the night before watching a documentary about the global coffee trade. The film, called *Black Gold,* focused on the Ethiopian coffee trade, and it was maddening. In demonstrating how commodity pricing of coffee put a low ceiling on what farmers could get for their coffee, it showed how much work there was to do in leveling the playing field for producers.

But there was one inspiring man in the film, an Ethiopian named Tadesse Meskela, who was on a crusade to change this paradigm. Meskela had managed to organize thousands of Ethiopian farmers and, by working in the specialty coffee export market, had significantly raised their per-kilo prices. But for every thousand farmers he could help, there were ten thousand more unable to compete, and who lived in poverty. In Ethiopia, the pay for a coffee worker was about one dollar a day. Mokhtar knew already that in Yemen, conditions were far better, that wages were more like ten dollars a day. Ethiopia was in a difficult spot, having more coffee, thus less scarcity, a reputation for uneven quality and unreliable delivery. But in the film, Meskela was both fierce and eloquent. He traveled the world representing Ethiopian farmers, and his successes were many, in the marketplace and in the realm of hearts and minds. He had opened a school and a hospital for his farmers.

And at the conference, there he was, a few feet away.

"That's Tadesse Meskela," Mokhtar told Rafik.

Rafik had no idea who Tadesse Meskela was.

"I'm going to introduce myself," Mokhtar said.

Rafik did not care.

Mokhtar was shy, though, and Meskela was never alone. Finally, at lunch, Mokhtar saw Meskela eating with two other Ethiopians. When Mokhtar approached, Meskela looked up, surprised. Besides the Ethiopians, Mokhtar was the only nonwhite person there.

"Hello sir," Mokhtar said to Meskela.

"Where are you from?" Meskela asked.

"I'm from just across the river from you. Yemen."

"Oh Yemen!" Meskela said. "I love Yemeni people."

And they were off. So many similarities in their challenges and opportunities, they agreed. They talked quality, and supply chain, best practices and plans. Meskela told Mokhtar he'd been to Yemen for a conference.

"Was that the Arabica Naturals Conference in Sana'a?" Mokhtar asked.

Meskela was impressed. They talked about how similar Harar, Ethiopia, was to Yemen. In Harar, they too were having trouble with qat, and were struggling to convince farmers to replace the qat with coffee.

"Whatever you do, help your farmers," Meskela said. By this time they were holding hands—a normal thing for men in Yemen and Ethiopia. "If you're in it for the money," Meskela said, "you won't last."

He gave Mokhtar his business card, which was oddly three times larger than any business card Mokhtar had ever seen.

"Visit me if you ever come to Ethiopia," Meskela said.

The man running the conference was Willem Boot, a charismatic Dutchman in his early fifties who knew everyone in the room. It turned out he'd coauthored a report on the state of Yemeni coffee, to be published later that year. The report was commissioned by the Coffee Quality Institute (CQI) and funded by the United States Agency for International Development (USAID). Mokhtar elbowed Rafik. At an Ethiopian coffee conference in Los Angeles, they had happened upon the man best situated to help him with Yemeni coffee.

In between conference sessions and cuppings, Mokhtar managed to get a few minutes alone with Willem Boot, and because he wanted to give Willem the idea that he knew what he was talking about, but because he talked so quickly, and because he didn't yet know what he was talking about, he sprayed Willem Boot with a random mix of words and phrases that had some passing connection to reality. "I want to radically improve and streamline the supply chain to the varietal farmer of the origin of the natural process of the . . ." After a while he decided to stop talking.

Boot looked at him with compassion. It was obvious to both of them that Mokhtar had fire but needed firsthand knowledge; everything he knew he'd gotten secondhand. Boot let him know that his company, Boot Coffee, offered consulting services to would-be importers and exporters, and also Q-grading classes.

"Okay," Mokhtar said, though he didn't know what a Q-grading course was.

There were fees associated with any of the services or courses, Boot added, but they were well within reach for anyone serious about starting a company.

"Right," Mokhtar said, though he had no money or access to it.

Mokhtar took Boot's e-mail address and said he'd follow up.

Boot didn't know what to think—was this a very young dilettante, some kind of hustler, or the genuine article? And how had he gotten into his conference for Ethiopian growers and American buyers? This seemed like the kind of thing Graciano Cruz would have arranged. He made a mental note to ask Graciano about this man named Mokhtar. In the meantime, he didn't expect Mokhtar to follow up.

But he did, the next day. Mokhtar said he wanted to hire Boot as a consultant. The only catch, Mokhtar said, was that he didn't know if he could get all the way to Holland. Mokhtar assumed Boot Coffee, run by a Dutchman, was in Amsterdam.

"Why would you go to Holland?" Boot said. "I live in Mill Valley."

CHAPTER XVII

STEALING COFFEE BACK
FROM THE DUTCH

MILL VALLEY WAS JUST north of San Francisco, but Mokhtar had
never been to Mill Valley. Mill Valley was in Marin, a county exotic
and unknown to Mokhtar, even though it was only ten miles from the
Tenderloin. It was green and overgrown, under the shadow of Mount
Tamalpais, an evergreen peak rising twenty-five hundred feet above
the Pacific coast. Willem and Catherine Boot's house stood on a winding road just off Miller Avenue, one of the town's main thoroughfares,
but it seemed very much apart from the world. A two-story adobe fes-
tooned with wisteria, bamboo and wild roses, their home could have
been in Tuscany or Greece.

That first day, Mokhtar brought Omar with him. After the Los
Angeles conference, Mokhtar called him and outlined the possibilities of taking what he'd learned about the Ethiopian coffee trade
and applying it to Yemen. Omar was interested. He wanted to meet
Willem, too. So the three of them sat outside at a long sturdy table,
the kind found in Italian movies where generations eat cheese and
prosciutto and children scamper underfoot. In dappled light, Mokhtar
and Omar got to know Willem, a man born into coffee.

His father, Jacob, had been one of the first Europeans involved in coffee's second wave, long before the popularization of specialty coffee. In the 1970s, when mass manufacturing was diminishing the overall quality of coffee made available to the public, Jacob Boot was trying to get the Dutch hooked on home roasting and single-cup brewing. He had been regional director for a Dutch roaster called Neuteboom, and as a child Willem visited the roastery often, swimming his hands in barrels of green beans. But on the side Jacob had harbored dreams of making and selling a home roaster, so people could control and appreciate the process in their own kitchens. He had invented, and gambled his life savings on, a machine called the Golden Coffee Box. He sold his house and opened a hybrid business where, in a rented storefront, customers could learn about high-quality beans, and a manufacturing arm, where he built and sold the Golden Coffee Boxes.

This was the seventies, a period when Europeans and Americans preferred their food fast and cheap. Jacob wanted people to slow down, to care about the origin of their comestibles. His business, based in Baarn, a town in Utrecht province, was not a runaway success, but sales of the Golden Coffee Boxes, and the coffee Jacob roasted, were strong enough that he could make a living. Eventually, when Willem was fourteen, he joined his father in the business and learned how to match every customer to specific flavor profiles. An artistic customer might prefer the brightness of a Kenyan coffee. A smoker might need something stronger, maybe a Sumatran bean. An older client might prefer a mellow Mexican Maragogype varietal.

The business was small and the storefront was never too busy, and Jacob liked it that way. He had time to talk with customers, to hear their stories, to find ways to pair them with the right beans. Jacob's attention to detail, his passion for every minute fact and nuance of coffee, was limitless. Willem watched his father drive importers to distraction with his questions about the origins of the beans he roasted. If the beans were from Java, he wanted to know where, which farm, what was the elevation, who were the farmers. Most of the importers had no idea.

Year by year, though, Jacob grew more tired, less able to do the day-to-day, and Willem took over more and more of the roasting. Jacob taught him what he knew, and Willem had the time and machinery to experiment, to gain a knowledge so extensive that after college, he went to work for the American subsidiary of Probat-Werke, a German maker of roasting machines. The company stationed him in California, and there he met his wife Catherine, a native of Napa Valley. Eventually they moved to Mill Valley and set up Boot Coffee, a roastery and educational institute where coffee growers, roasters, academics and other aspirants could learn. Willem became a consultant, training roasters and growers, and emerged as a sought-after judge in coffee competitions around the world, from Central America to Ethiopia and Papua New Guinea.

That first day in Mill Valley, Willem already had a plan for Mokhtar.

"If you're serious about this," Willem said, "you should become a Q grader. Then you should go to the Specialty Coffee Association of America conference. That's in Seattle. Then you should go to Yemen."

"Okay," Mokhtar said, already calculating the costs.

"If you want to improve coffee exported from Yemen," Willem

said, "you have to know what's good and what isn't. And as far as I know, you'd be the first Q grader for arabica coffee who's an actual Arab."

"Great," Mokhtar said, still having no idea what a Q grader was.

"What's that course cost?" he asked.

The course cost two thousand dollars. And to engage Willem as a consultant would require a retainer of five thousand. The irony was not lost on Mokhtar and Omar that two Yemenis—direct heirs to the ancient trade of coffee—were being asked to pay to learn about the business from a Dutchman.

After the meeting, Mokhtar and Omar got into their cars, and Mokhtar followed Omar a few miles west, onto Highway 1. They stopped at a turnout overlooking the Pacific and performed their noontime prayers. When they rose, Mokhtar said it was too much money. He couldn't ask Omar for it. Omar had already loaned him three thousand dollars after he'd lost the satchel. Still, he took out his checkbook and wrote Mokhtar a check for five thousand dollars. A line of credit, he called it.

Nothing like that had ever happened to Mokhtar. His family and his occasional boss or teacher had been encouraging when he'd conjured an idea or even a plan, but no one had ever stepped up like this. He'd never seen that much money in one place. He took the check, got into his car, watched Omar drive away and, alone and overwhelmed, he let himself cry until he couldn't anymore.

CHAPTER XVIII

THE APPRENTICES

MOKHTAR FORMALLY ENGAGED WILLEM and Boot Coffee,
but the relationship was not quite formal. The five-thousand-dollar
retainer really was just a way in the door, and once inside, Mokhtar
did what he'd done at Blue Bottle, insinuating himself so subtly but
thoroughly into the operation that within a week, half the staff at
Boot Coffee wasn't sure if he worked there or not.

Every day, Mokhtar left early in the morning, beating traffic
to drive an hour north, then over the Richmond Bridge and south
again into Mill Valley, always arriving when Boot Coffee opened. He
did anything asked of him. He ran errands to Safeway and Whole
Foods. He cleaned the machines and swept the floor. He watched.
He listened. Willem and Catherine traveled frequently, to Panama,
to Nicaragua, to Europe, and when they were gone, Willem left the
operation in the care of his young staff, Stephen Ezell and Jodi Wieser
and Marlee Benefield.

Stephen was not much older than Mokhtar, and had accomplished
the same sort of self-invention Mokhtar was attempting. Stephen was
a Florida native who'd studied philosophy in college, had bartended

and played in bands until he decided to follow his brother to the Bay Area. Stephen was perusing the job listings one day when he circled three intriguing options. One was in wood refinishing, the other in biohazard removal, and the last offered a "coffee opportunity in Mill Valley." Stephen had worked briefly at a Starbucks in 1999, so he sent his résumé.

And though he'd been at Boot less than a year, Stephen, with his rust-colored beard and deliberate manner, had the air of an Old World apprentice. He did most of the roasting at Boot, making it a delicate blend of art and science—with a degree of precision and instinct that Mokhtar wasn't sure he'd ever be able to replicate.

If Stephen was the young apprentice, Jodi Wieser was the journeywoman—a few years ahead of Stephen in her training. Young and thin, with glasses and strawberry blond hair, she was a center of calm competence. She'd grown up in Dallas, and in high school had worked as a barista at a high-end café where a cup of coffee went for $4—and that was 1996. After college, she'd moved to Africa to do NGO work, first in Mali, then Ivory Coast. She moved back to the United States, got a master's in intercultural studies and returned to Africa, this time Uganda, where she helped establish a nonprofit, Fount of Mercy, that worked with AIDS widows, orphans and former child soldiers.

She was joined there by Marlee Benefield, a friend from grad school with the youthful face and sunny attitude of a summer-camp counselor. When Jodi moved back to the U.S., she was looking through Craigslist for jobs with an international angle. She saw an ad Willem had placed. *Must love coffee,* it said. She joined Boot Coffee in 2008 and became a Q grader in 2010. Marlee followed shortly thereafter and began selling roasters and teaching courses in roasting. Together, Jodi, Marlee and Stephen were, compared to Willem, relatively recent arrivals, and all of them were guided by Willem's example—they were experts but not snobs, serious about the work without being overly serious about themselves.

The one thing Willem was adamant about, though, was that Mokhtar needed to become a Q grader. Mokhtar knew this, too, though he still didn't know precisely what that meant, and there was the small matter of his knowing not much about how coffee tasted. He hadn't advertised this to anyone, least of all anyone at Boot, but up until then Mokhtar had only had a few dozen cups of coffee in his life. There had been the espressos he'd had with Giuliano, the sips he'd had at Blue Bottle. What had interested him about coffee first had been the history, his pride in Yemen's central role in its cultivation and dissemination. At Boot, though, he slowed down to savor the taste and discern the varietals, the brewing methods and permutations. At Boot he could finally let his guard down, admitting all he didn't know.

He became familiar with the drinks, exotic twenty years ago, now standard. There was an espresso, which contained basically the same amount of beans as a cup of coffee, but finely ground and concentrated in a much smaller amount of water. There was a café au lait, half coffee and half steamed milk. There was a macchiato, a double shot of espresso topped with foam. There were the drinks borrowed from or based on drinks popular abroad. The espresso *romano* was a shot of espresso with a slice of lemon served on the side, not to be confused with the guillermo, one or two shots of espresso poured over slices of lime. Algeria gave the world the mazagran, a cold coffee drink made with coffee and ice, and sometimes rum or sugar or lemon, served in a tall glass. There were various coffees with cheese—cheese dunked in a cup of hot coffee and later eaten when the mixture softly congealed. In the Spanish-speaking world, there was *guarapo con queso*, made with Gouda or Edam. Among the Swedes it was called *kaffeost* and employed the Finnish cheese *leipäjuusto*. There was iced coffee and cold-brew coffee. There was Thailand's black tie—a mixture of black tea (chilled), sugar, and condensed milk, crushed tamarind, star anise, orange-blossom water, and a double shot of espresso. There was Irish coffee (with whiskey), English coffee (with gin), and calypso coffee (with Kahlúa and rum). In Senegal there was *café touba,* where the beans were mixed with *selim* and other spices during roasting, and sugar was added to the hot coffee—making a very sweet and aromatic drink. In Australia there was the ice shot, a shot of espresso dumped into a latté glass filled with ice.

At the fringes were some very strange things, none odder than *kopi luwak,* otherwise known as civet coffee. Coffee had been grown in Sumatra for 150 years, but only recently was it discovered that the civet, a catlike mammal indigenous to the island, was something of a coffee connoisseur. The civet was expert at picking the ripest cherries to eat, and afterward, its feces were found to have done all the processing work usually requiring man and machine and much water. That is, while the coffee cherry passed through the civet's digestive system, the skin and pulp and mucilage were removed, leaving only the beans, which the civet couldn't digest. Someone got the idea to take these beans and separate them from the feces—to pick up civet feces, and pick coffee beans from the feces, and to roast and grind and drink these beans. Something in the civet's digestive tract gave the coffee an unusual and strangely appealing taste—musky and smooth. *Kopi luwak* became popular, and its purveyors were able to demand a premium for it. Willem was not impressed. He liked to repeat an expression coined by George Howell, a well-known coffee roaster. "Coffee from assholes, for assholes," he said.

Willem, Stephen, Jodi and Marlee welcomed Mokhtar's help. Boot Coffee was so overrun with visitors that having one more hanger-on didn't make much difference. A bewildering parade of nationalities came through for classes, cuppings and consultations: Koreans and Uzbekis, Japanese and Croatians and Russians. Plenty of Germans, plenty of Dutch, some French, some Canadians, a few Malaysians, lots of Chinese and Australians. They gathered at the round table, cupping and noting and drawing on the chalkboard. The sessions lasted hours, and would be decidedly unpretentious. People even laughed. There was a looseness and an openness that would have surprised anyone who assumed that a cupping, where coffees were rated and discussed, would be insufferable. Still, the sentences spoken could be rarefied.

"I taste classical jasmine and rose."

"I found it a little brassy."

"I picked up some Fuji apple."

"I tasted a whole flowerpot."

"The promise of aromatics was so strong that I was disappointed in the aftertaste."

The cupping and scoring done, often it was up to Mokhtar to tidy up, to pour the extra coffee down the drain and wipe the counters. He liked to stay late, to be alone with the cooling coffee, with the idle machines with their gauges and knobs, the beans in their carefully labeled containers—the space a hybrid of kitchen, chemistry lab and furnace room. With everyone gone, Mokhtar would take up a clean spoon and circle the cups, reading the notes and scores attached to each.

He needed to taste what the others were tasting.

CHAPTER XIX

PASSING THE Q

HE WASN'T A Q grader, and knew he had to be. A Q grader was essentially an expert on the quality of arabica coffee and uniquely qualified to score it. An R grader was an expert on robusta, but that was considered a far-less-prestigious thing to be. A Q grader has completed intensive coursework and has passed a rigorous test to prove he or she can differentiate between bad coffee and good coffee, between good coffee and superior coffee. Being a Q grader is something akin to what a sommelier is to wine, a grand master is to chess. Like so much of third-wave concentration on quality and expertise, the Q-grading program was very new, established in 2004. Ten years later, there were still only two thousand Q graders in the world. And Willem was correct: among those two thousand, not one was an Arab. This seemed like the kind of obvious challenge that Mokhtar was meant to overcome. A vision gripped him, of returning to Yemen, arriving in Sana'a and striding through the country as the world's first Arab Q grader. An important man.

The class was two grand, and he couldn't ask Omar for more money. His mind searched for other donors and arrived at his uncle

113

Muteb, his father's younger brother. Muteb, an entrepreneur, lived in Modesto, where he ran the family grocery business, expanding Hamood's operations to a string of stores along Highways 99 and 5. Muteb would understand.

Mokhtar didn't tell Muteb why he was coming. He only said he had some news. When he arrived, Mokhtar set up all his gear as Muteb, his wife Layla and their seven kids, all under fourteen, watched, baffled. Mokhtar had brought three varietals from Ethiopia, his Chemex set, a small digital scale, a gooseneck kettle, a coffee grinder and a popcorn maker, which he planned to use to roast the beans the Ethiopian way—flat on a pan.

He weighed the coffee beans and roasted them, the exuberant aroma filling the room. When they had cooled, he took out his grinder.

"What are you doing?" one of the kids asked.

"I'm grinding the beans," he said. "Otherwise it's just a bean. I grind the beans into a powder so they can dissolve in water. And it's key that we use a medium-coarse grind."

"What does that mean, 'medium-coarse grind'?" they asked.

"As opposed to fine or coarse. When we're using a Chemex, we favor a medium-coarse grind." Stephen had taught him this.

"What's a Chemex?"

Mokhtar showed them the glass vessel, a clear carafe, about ten inches high, cinched tight in the middle, giving it the look of an hourglass with an open top. "This is where I'll pour the coffee. I'll pour it into the top, and the filters will catch the grounds but let the coffee go through."

"Is that the filter?"

"A special kind," he said. It was simple, though, just a square piece of filter paper, really, but the handling of it was key. He took the square piece of filter paper and folded it to a fourth of its size, then shaped it into a funnel—it was crucial that one side was three-ply and the other one-ply. Mokhtar ran the funnel under the faucet, to make it moist, and then inserted it onto the top of the glass vessel.

"You gonna pour the water now?" the kids asked. They were suddenly impatient.

"The water has to be the right temperature," he said. "Let's see how it's doing."

The water had been boiling in the small gooseneck kettle, the ideal temperature between 195 and 205 degrees. The pot sat on a small dais with a digital readout.

"You tell me when the water's at two hundred and two," he said.

Using a tablespoon, Mokhtar scooped the ground coffee into the funnel and spread it evenly.

"It's at two hundred and two!" one of the kids reported.

"Now watch," Mokhtar said as he poured the water into the funnel in three distinct stages. First, he poured water onto the filter to soak the paper and to flush out any taste of paper. Then he poured just enough water to soak the grounds.

"Why are you waiting?" they asked.

"I have to wait forty-five seconds now," he said. "This is called the bloom period, when the coffee's gases are first released."

They waited the forty-five seconds, and then Mokhtar carefully poured the rest of the water in, circling it around the grounds until the gooseneck was spent.

"Now what?"

"Now we watch the coffee drip down," he said.

The water soaked up the grounds, emulsified the coffee, and dropped through the filter and into the lower half of the Chemex. Mokhtar had made enough for three cups, and when it was ready, he removed the filter, threw it into the compost, and poured cups for Muteb, Layla and himself. Muteb and Layla politely sipped, and as they did, Mokhtar talked about the history of Yemeni coffee and hinted at his plans to get involved in importing beans from their ancestral home. At the end of the presentation, though, Muteb and Layla didn't understand the connection between the elaborate coffee presentation they'd just witnessed and what Mokhtar was going to do with his life. Their nephew was twenty-five years old and had no job and no college degree. The steadiest work he'd had was as a door-man. Muteb didn't see how pouring coffee was going to do much to advance his life goals.

"Aren't you supposed to be in law school?" he asked.

Back at Boot Coffee, Willem had conjured a plan. He had cowritten his Coffee Quality Institute report on Yemeni coffee with another expert, Camilo Sánchez, and Camilo and Willem thought that with CQI funding, they could put together a trip to Sana'a. They would gather Yemeni growers and international buyers, and maybe forge some connections to help the country's struggling coffee industry.

Mokhtar would go with them and serve as translator and cultural bridge, but most important, after the conference, he and Willem and Camilo would travel around the country on what Willem was calling a coffee caravan. They would take an SUV or two into the country's coffee producing regions, meeting the farmers, roasting beans and cupping coffee. They would identify which regions were producing high-quality cherries, and where they might be able to help farmers with education and potential partnerships. All along, they'd have a grand time.

This would be the culmination of Willem's consulting services, and then it would be up to Mokhtar to continue the work in Yemen.

"But you need to be a Q grader first," Willem said.

Now it was April 2014, and they planned to leave in May. Boot Coffee was holding one Q-grading class before the conference in Sana'a, so Mokhtar had no choice but to go back to Omar for more help. It put him in a position of extraordinary debt to one man, but Omar didn't hesitate—he put up the fee for the class, and Mokhtar enrolled.

The instructor was not some enigmatic professor from a faraway land but Jodi Wieser, whom he'd been working next to for months.

"I'm nervous," she told the class. She'd taught dozens of Q-grading sessions, but never as lead instructor, and the pressure on an instructor was extraordinary, given the expense of the course and the great distances the students had traveled to take it.

They were a far-flung group, and much was riding on their success. There were two students from Mexico who ran Buna Café Rico in Mexico City. They seemed to be the most experienced and most confident. The other two students embodied the kind of high-stakes pressure so many of the students felt. One student, a woman in her thirties, had taken the Q test twice and failed. If she didn't pass this time, she'd have to retake the entire class. And this time, she was pregnant. It was a factor Jodi hoped would help her—perhaps her increased olfactory sensitivities would be to her advantage. The last student was a Kuwaiti man who'd arrived a week early to take classes with Willem. During the coursework with Jodi, he seemed burdened. He had come very far for this, determined to be the first Q grader from Kuwait, just as Mokhtar wanted to be the first Yemeni.

Though there was sometimes a certain status seeking among some students—a way to be or seem elevated among their peers in the specialty coffee world—Jodi grounded the course in the real impact Q grading could have on producers, farmers, on every part of the production chain. That was the original motivation for the Q course. It was instigated by the Coffee Quality Institute as a way to empower coffee growers. In many producing regions, particularly before the third wave, the farmers don't know much about their coffee. Very often they didn't drink it—especially in Yemen. Because they didn't know much about the quality of their own product, they were at the mercy of brokers and commodity pricing. But if farmers became expert Q graders, they would know what they had. If they had a great coffee, they could rate it, and find buyers who would pay far-higher prices for a high-scoring coffee than they would a coffee of unknown quality.

The CQI, then, was committed to leading as many farmers as possible into Q-grading courses. When the farmers, millers, exporters, roasters and retailers were all having a common conversation about the same coffee, then real empowerment could happen. If a Rwandan producer knows how to improve his coffee, and can cup it and rate it, he can bring it to the 90s. On a scale that topped out at 100, anything in the 90s was considered extraordinary, and that farmer could transform his business from a low-wage commodity at the mercy of the world commodity market into a specialty business wherein he could work directly with roasters of his choosing.

This was the gist of Jodi's first-day speech. She told the story of going to a farm in Panama, and cupping with that farm's owner, who had become a Q grader a few years before. She and the farmer had cupped twelve local varietals, and on every single coffee, their scores were within one point of each other. Having a common international language to assess quality made for a powerful economic tool.

The course was hard, and the tests were harder. Only about 50 percent of those who took the Q test passed the first time. There were twenty-two parts to the test, some of which would seem, to the generalist or everyday coffee enthusiast, insane in their specificity.

The most accessible and concrete part was the General Knowledge exam—one hundred multiple-choice questions about coffee cultivation, harvesting, processing, grading, roasting and brewing. The rest of the tests, though, required a freakish level of sensory sensitivity.

There was the Olfactory Skills test, wherein the blindfolded student had to discern thirty-six different scents, including garden peas, maple, cooked beef, butter and tea rose.

The Cupping Skills test required the student to identify and rate various coffees, African and Asian, mild and strong, processed and naturals, and these ratings had to line up with precedent. If a coffee had been previously rated at 94, the student, testing it blind, had to score it within two points of the established rating.

For Triangulations, the student was given six sets of three cups. Two of the three cups were identical, and one was something else. The student had to identify the outlier. The Organic Acids Matching Pairs test started with eight sets of four cups of coffee. Two cups in each set were tainted with some kind of acid—phosphoric, malic, citric or acetic. The student had to be able to tell which cup had been altered, and by which acid. For the Sample Roast Identification test, the student started with four cups of brewed coffee and had to discern which was overroasted, which was underroasted, and which was perfect.

The Arabica Roasted Coffee Grading test required the student to take a one-hundred-gram sample of roasted beans and identify any quakers—undeveloped beans that didn't roast correctly—and also identify if the sample was of commercial, premium or specialty grade.

In his months at Boot, Mokhtar had been around the course as others had taken it. He'd cupped and watched dozens of cuppings. But when it came time to take the test, he felt like he was starting over. And Jodi was nervous for him. And for the Kuwaiti. And for the pregnant woman. Jodi wanted everyone to pass.

After the testing, the students were given their results. The two Mexican students passed easily. The pregnant woman passed. But the Kuwaiti didn't, and when it was time for Jodi to tell Mokhtar, she hesitated. She scratched her neck, stared at the floor. She couldn't look him in the eye.

"You didn't pass," she said. "You failed seven of the tests."

Mokhtar paused. He did the math.

"You mean I passed *fifteen* of the tests?" He was ecstatic. He hadn't scored that high on a test since junior high.

But there was no time to retake the test before the trip to Yemen. He would return as a young man with an idea, but not yet the Important

Man he'd imagined he'd be when he made his return. Only now did he tell his parents what he'd been up to. He showed them a picture of his coffee equipment, the same apparatus he'd brought to Muteb's house.

They did not see coffee as a serious use of his time.

When he told his brother Wallead he was going back to Yemen, Wallead said, "Really? Who died?"

BOOK III

CHAPTER XX

HAMOOD AND HUBAYSHI

"DON'T TELL ANYONE WHY you're here."

That was his grandfather Hamood's advice.

"Tell them you're doing a report for college."

"But I'm not in college," Mokhtar said.

"They don't know that," Hamood said.

The risk in coming back to Yemen, his grandfather explained, was inviting interference. Relatives would want to get involved. Or they would give advice. Or friends of relatives would want to get involved or give advice. There would be random people attaching themselves to the idea, getting in the way, reshaping it or, worst of all, trying to do the same thing, only quicker and cheaper.

Mokhtar booked his ticket to Sana'a, telling everyone but his grandfather that he was doing a research paper about the history of coffee in Yemen. There was nothing less intriguing to everyone he knew in Yemen than the idea of a college student doing research. He would be left utterly alone.

And a student was presumed to be broke, so there wouldn't be that problem, either. American Yemenis were known to come back from the U.S. with suitcases full of money, ready to throw it around. Mokhtar had to keep a low profile, appear young and unemployed, and stay invisible. He had a highly specific plan to execute—the first part of the plan was to transform Yemeni coffee—and it involved his grandfather, the province of Ibb, and Willem Boot.

He would first go to Ibb, three hours south of the capital, where Hamood lived. Hamood would introduce him to those he knew in the region's agricultural businesses. Then, in a week or so, Mokhtar needed to be back in Sana'a, to attend the Coffee Quality Institute workshop in Sana'a with Willem and Camilo. Then the coffee caravan.

The caravan was everything. Mokhtar saw this as the ideal way to familiarize himself with coffee in Yemen. He would be with Willem and Camilo, two of the world's foremost experts on coffee quality. Mokhtar would watch them, learn from them, and meanwhile he would act as a bridge to the Yemeni farmers, speaking their language, sharing their history. The coffee caravan would give Mokhtar introductions to all the Yemeni coffee farmers, and he would be traveling with an entourage that would impress the farmers and give him the kind of standing that would launch his Yemeni coffee career.

Meanwhile, the U.S. State Department was advising all travelers to avoid Yemen. But Willem was traveling at the behest of the U.S. government, so how bad could it be?

Mokhtar knew about the Houthis, a rebel group from the north of Yemen. They had waged an insurgency against the government of Ali Abdullah Saleh for six years before the Arab Spring. They took part in antigovernment protests in 2011, but after Saleh was ousted and Abed Rabbo Mansour Hadi ascended to the presidency, the Houthis remained critical of the government and were presumed to be aligned with Iran. Meanwhile Saleh, forced out after the Arab Spring, now had designs on returning to power. Then there was al-Qaeda in the Arabian Peninsula, which had gained strength during the power vacuum in the wake of the Arab Spring and was considered the most dangerous al-Qaeda franchise in the world. To Mokhtar it all seemed to be part of the neverending political churning of Yemen, and for the time being none of it had anything to do with his immediate aims. He needed to get to Ibb.

A distant cousin, recently married in the capital, picked him up in one of Hamood's cars, his new wife in the passenger seat, and Mokhtar and the newlyweds made the three-hour drive to the home of Mokhtar's grandparents, Hamood and Zafaran. Zafaran was, at the time, back in the U.S., living in California, so Mokhtar would be spending time alone with Hamood, the John Wayne of Yemen.

When the newlyweds dropped Mokhtar off, Hamood greeted him from the ornate door of his estate. Mokhtar could see his grandfather had aged. His back hunched a bit more, and he leaned heavier on one of the hand-carved canes he favored. He and Mokhtar walked the grounds of Hamood's compound, past the guava and fig trees, under a blue sky.

"You told no one what you're doing?" Hamood asked.

"No one."

"Good."

They came to the row of coffee trees hugging the wall of Hamood's compound. "You remember these?" Hamood said.

Mokhtar touched the glossy leaves. He remembered. When he'd lived with Hamood and Zafaran as a teenager, he'd seen these plants every day but had never realized they were coffee. He'd used the cherries as projectiles, and occasionally chewed on the fruit's outer layers, but now, for the first time, after his eighteen months of research, he was actually touching a coffee plant and knowing it was a coffee plant.

The leaves were surprisingly firm and lustrous. Their edges were wavy, their surfaces rippled. They were strong leaves, colored a rich Kelly green, and under each leaf array, cherries took shelter. The fruits were baffling in their variety. In any group of fifteen cherries there were fifteen stages of readiness. Some were a bright green, others chartreuse, some transitioning to orange, and then a few fuchsia and, finally, three or four fully red. He plucked a bright red cherry from the tree, feeling the resistance from the stem—the tree didn't willingly give up its fruit.

The labor intensiveness, which Mokhtar had read about and knew from Willem and Jodi and Stephen and Camilo and Tadesse, now became alarmingly real to him. To approach one of these trees, then isolate a fruit cluster, then examine fifteen fruits from that cluster, then pull from the cluster only the three or four cherries that were ready that day, each one resisting just a bit—it would take time. It was like shopping for fruit at a market, taking the time with each apple or melon, looking for bruises, examining the color. A picker doing this for every cherry on every tree—that was significant work. To do it well would take both a discerning eye and real physical stamina.

Mokhtar sat down next to Hamood, who had rested his cane against the low stone wall.

"If you're going to do this," Hamood said, "you should meet Hubayshi." Hifdih Allah al-Hubayshi was the biggest local trader in coffee, a dominant force in the business for fifty years. He had billions of riyals—millions of dollars—under his control, Hamood said, but he wore it lightly. He was highly ethical and fair in what was considered a cutthroat business.

"So you know him?" Mokhtar asked.

"I've never met him," Hamood said.

Hubayshi did not look like a wealthy man. Hamood had sent Mokhtar alone to meet him, giving Mokhtar the address but not much guidance. When Mokhtar arrived, he found a man dressed in tattered clothes, operating out of a small storefront in downtown Ibb. He was likely the same age as Hamood, but looked far older. Mokhtar introduced himself, expecting Hubayshi would be impressed that he was Hamood's grandson, that they would immediately begin a partnership. But he got little respect or attention. Hubayshi was brusque and wary.

"You're a student?" he asked.

"Yes sir," Mokhtar said.

Hubayshi didn't seem to believe it. Their meeting ended quickly, and Mokhtar walked back to Hamood's house feeling rattled. The most influential man in the regional coffee business didn't want anything to do with him. And what did it mean that the most successful man in Yemeni coffee looked like a pauper?

For the time being, it didn't matter. Mokhtar had more immediate concerns. Willem was coming to Yemen in a few days, and Mokhtar had to be ready. The Coffee Quality Institute was putting on the conference with participation from USAID, with the gathering meant to strengthen connections between Yemeni coffee farmers, brokers and international traders. Mokhtar was to be on a panel with Willem and Camilo, and through Willem, Mokhtar would meet the players in Yemeni coffee, in international distribution and afterward there would be the coffee caravan. Willem had his own ideas of regions he wanted to visit, but Mokhtar knew he would be crucial in gaining access to certain tribal areas. He pictured the journey vividly, the three of them traveling the hills and valleys, meeting farmers and collectives, picking and roasting and cupping, laying the groundwork for Mokhtar's future in the business. But first Mokhtar needed to establish himself in the capital, and his grandfather Hamood didn't have a place there. Mokhtar's mother suggested her uncle Mohamed.

Mohamed was from Ibb, had worked for many years in Saudi Arabia as an electrician, and had recently retired in Sana'a. Of modest means, he and his wife Kenza lived in a building owned by Kenza's brothers, Taha and Yasir, and relied on income sent home by their son Akram, a janitor at the Contemporary Jewish Museum in San Francisco. His remittances paid the bills for Mohamed and Kenza and their three daughters and three younger sons still at home. Their house was in the center of Sana'a, so through Mokhtar's mother it was negotiated that Mokhtar would come to stay for a time, even though they had little room and Mokhtar would have to sleep on the floor.

Which he did. He arrived in early May and quickly made an arrangement where he'd unroll a blanket at night, sleep in the corner of the living room and, in morning, roll it up, hide his clothes in a corner under a chair, and in general try to be invisible. In exchange, he'd find ways to contribute to the household without causing his great-aunt and great-uncle to lose face. Instead of giving them money outright, Mokhtar bought groceries and household essentials, cleaned and helped with the girls' homework. At meals, Mokhtar and Mohamed talked politics. Everyone in Yemen talked politics; there was never a dearth of new developments, and Mohamed had seen political violence firsthand.

When he was growing up in Ibb, Mohamed had seen heavy fighting in the 1970s and '80s between the Yemeni government and those who wanted the region to be a socialist state. The socialists, benefiting from significant military and financial support from the Soviet Union, tried to purge the region of tribalism, and made it a point of strategy to eliminate local chiefs. One of their targets was Shaykh Mohamed Nashir al-Khanshali, tribal leader of al-Dakh in Ibb province—and, improbably enough, the brother of Mokhtar's grandfather Hamood. In 1986, al-Khanshali was driving his car when he was struck and killed by a rocket fired from a bazooka. Mohamed saw it happen. He pulled al-Khanshali's charred body from the vehicle.

When Mokhtar arrived in May of 2014, the Marxists were long gone, but intertribal warring in Yemen was again in full swing. Historically, Yemen, when not being invaded or colonized by outside powers, from the Ottomans to the British, was fighting itself. It wasn't until 1990 that Yemen had become the Arabian Peninsula's first multiparty parliamentary democracy. In 1993, elections were held, and in 1999, Field Marshal Ali Abdullah Saleh was elected president of the newly unified country. He was not popular for long, and the Arab Spring swept Yemen up in its dreams of a more democratic and equitable Middle East. Under pressure from within Yemen and from the international community, Saleh eventually resigned. He was replaced by Abed Rabbo Mansour Hadi, but by then the Arab Spring's yearlong power vacuum had emboldened insurgent movements. There were the Houthis, a rebel group named after its leader, Husseyn al-Houthi, who were dissatisfied with the leadership in Sana'a—who historically ignored their region, they felt—and had been staging raids and seizing land in the north. In the south, with Aden as its capital, there was talk of secession.

Meanwhile there was the growing presence and threat of al-Qaeda in Yemen, known as al-Qaeda in the Arabian Peninsula (AQAP). Al-Qaeda had been operating in Yemen for twenty-two years, beginning in 1992, when it bombed a hotel in Aden commonly used by marines; that killed two people. There was the 2000 attack, off Aden's coast, on the USS *Cole,* which took seventeen lives. In 2007 eight Spanish tourists and two Yemeni drivers were killed by a car bomb in the province of Marib, and a year later, another twelve civilians were killed by a car bomb outside the U.S. embassy—Benghazi before Benghazi. In 2009, a suicide bomber from Yemen was killed in Jeddah while trying to assassinate Saudi Arabia's top counterterrorism official. (The would-be assassin had detonated a bomb hidden in his anus, killing himself but only injuring the Saudi minister.) In 2011, AQAP took control of Zinjibar, a city in Yemen's south. In 2012, they coordinated a suicide attack near the presidential palace in Sana'a, killing more than one hundred Yemeni soldiers.

The United States, with Yemeni cooperation, had for years been targeting AQAP with drone strikes, which for Yemenis had become a fact of life. In April 2014, at least four drone strikes had been confirmed, killing anywhere between thirty-seven and fifty-five people, including between four and ten civilians, depending on whose report one was reading. On April 19, just a few weeks before Mokhtar arrived, CIA drones struck a truck carrying suspected militants, killing ten of them but also killing three laborers who happened to be nearby.

Still, in the spring of 2014, there was reason to be cautiously optimistic. President Hadi had just overseen the National Dialogue Conference, and after ten months of discussions, the delegates agreed on the basic provisions of a new constitution. Shortly after, a presidential panel approved a plan to make Yemen a federation of six regions.

Whether this would assuage the Houthi rebels in the north was unclear. For now, Mokhtar needed to worry about Willem, who was arriving in a few days from California. Assessing the risk was difficult and highly subjective. The capital wasn't considered a particularly safe place for Westerners, but most international embassies were still operating in Sana'a, and there were still thousands of tourists and foreign workers all over Yemen. There were commercial flights going in and out of the capital, so that pointed to at least some level of confidence in the relative stability of the country. But then again, the State Department warnings about travel to Yemen were dire. How and why the USAID conference was going on as planned was not entirely clear.

Then there was the recent warning, issued by video from AQAP's leader Nasser al-Wuhayshi, that they would hunt down "crusaders" from countries such as the United States, England and France. In the week before the conference, there had been attempted kidnappings of German and Russian nationals in Sana'a. On May 5, the day before Willem arrived, a French security contractor, guarding the European Union delegation in Yemen, was killed and another contractor wounded when gunmen opened fire on their car in the diplomatic area of the capital. The same day, a security officer was killed by two gunmen on a motorcycle outside the Defense Ministry's linguistics institute. In all, fifty-three people had been assassinated in March and April.

Willem Boot had been in dicey places before, and Camilo, as a Colombian, was not unfamiliar with dangerous settings or unstable governments. But they couldn't have prepared themselves for what they encountered in Sana'a. Before leaving the United States, Willem had received an e-mail from GardaWorld, the firm handling security for the conference. In all capital letters, the e-mail insisted that the recipient open the e-mail, print its attachments, sign and return all applicable forms. Included in the e-mail was a twenty-six-page Yemen security handbook, containing highly detailed and alarming security instructions covering a wide range of possible events—a terrorist attack, hostile crowd situations, potential mail bombings and kidnapping, which the guide called extortive assault, or short-term kidnapping. One document, titled Isolation Preparation (ISOPREP), requested a two-page list of Willem's personal characteristics, handwriting and next of kin, so authorities could identify him in the event of abduction. The handbook said, "There is no need to overreact to security problems, merely to keep one step ahead of the opposition. Applying a small percentage of time and effort to one's personal security can positively deter the terrorist and deflect his interest towards softer and more accessible targets."

When Willem landed in Sana'a, he was met by an Irish security contractor who said he'd be taking care of his welfare while in Yemen.

He guided Willem to two large armored SUVs parked outside the airport, and after that, Willem never saw him again.

In one SUV, there was a driver and another security contractor, armed with a machine gun. The second SUV followed, carrying three contractors, all heavily armed. The contractor riding with Willem gave him an envelope, and in the envelope was a cell phone and instructions to turn it on and call the number listed, for a man named Khaled. Willem tried to turn the phone on, but it didn't work.

The SUVs sped through the city, stopping at no checkpoints. At the hotel, armed guards stood outside the gate and in front of the lobby. Willem assumed he was safe, so he took his bags up to his room, unpacked and, with nothing better to do, began brewing coffee. He'd brought his Chemex kit and two large French presses.

That night Mokhtar walked out of his uncle's house, got into a taxi and directed the driver to the hotel. It was simple for him—there were no guards, no cell phones in envelopes, no security contractors from Ireland. When Mokhtar and Willem met in the hotel lobby, and Willem related the story of getting to the hotel, something subtle shifted in their relationship. Willem was his teacher, but now Willem was in Yemen. He needed Mokhtar as much as Mokhtar needed him.

They decided to do a test run—just a short trip out for dinner, to gauge the security risks for a few foreigners traveling within the city. Mokhtar arranged to have his family's beige Lexus SUV driven by his grandfather's driver, Samir. Samir arrived at the hotel in the early evening and Willem, Camilo and Mokhtar got in. Within a few blocks, they were stopped at a checkpoint. The SUV's tinted windows obscured the presence of Willem and Camilo in the backseat, so Samir and Mokhtar did the talking and got them waved through. It worked at the second checkpoint, too.

The third checkpoint was different. The soldiers there seemed to be an irregular unit. They wore different clothing and scarves and almost immediately they were shining flashlights through the windows, discovering Willem and Camilo.

"Who are these people?" the soldiers demanded. "What are they doing here?"

They ordered all the windows open, all the doors.

"We're just going out to eat," Mokhtar said.

But he'd lost control of the conversation. He was rattled, and Willem heard a change in his tone of voice. Mokhtar sounded tentative, unsure.

The soldiers, all chewing qat and seeming agitated, inspected Willem's and Camilo's passports and searched the car. Willem thought of kidnappings, that they'd be taken away and sold off. Mokhtar was thinking the same thing—not that he'd be kidnapped, but that his two friends and mentors, for whom he felt entirely responsible while in his country, were about to be made into currency or worse. Weeks ago, in Mill Valley, they'd all gone out to dinner—Mokhtar, Willem and Catherine, and their son Vincent. Catherine had expressed her fears about the trip, and Mokhtar could see in her eyes that she was gravely concerned. He reached across the table and took her hands and said, "Willem is now part of the al-Khanshali tribe, and we will protect him with our lives." She thanked Mokhtar, seeming reassured, and said, "Just make sure he doesn't bring home a second wife."

Mokhtar pleaded with the soldiers. "Please. I just want to show them true Yemeni food." Every time one of the soldiers looked at Willem or Camilo, though, Mokhtar assumed they were sizing up the two foreigners for resale value.

"We're really just going out to dinner," Mokhtar said, and even named the restaurant, noting how much better it was than the bland food at the hotel. "You guys can come with us. Meet us there. I'll buy you dinner."

Finally the soldiers relented. They let Mokhtar's group pass, and once they were at the restaurant, eating a fitful dinner, Mokhtar half expected at least one of the soldiers to show up. None did.

In good times and bad, kidnappings of foreigners were common in Yemen. In most cases the kidnappings were motivated by a tribe's desire for money or a prisoner swap, or for the Yemeni government's attention to their needs and demands. They would kidnap visitors from Europe and Asia and hold them in hopes of bringing awareness to flaws in the electrical grid in their region, for example. Almost without exception the captives were treated well and released unharmed. A year earlier, a Dutch couple, kidnapped outside their home in one of the safest neighborhoods in Sana'a, was held for six months and released unharmed. The couple praised the treatment they'd received and was careful to say how much they still loved Yemen. It was a bizarre but accepted cost of traveling in the country—the imminent possibility one would be kidnapped in order to solve an issue with regional infrastructure.

But the era of al-Qaeda brought a distinct change. In 2009, the mutilated bodies of two German nurses and a South Korean teacher were found, and these and other incidents underlined the marked difference between the Yemeni way and the way of al-Qaeda. All this was on Mokhtar's mind. He couldn't put Willem in danger, or allow Willem to put himself in any danger. From the restaurant, they went home shaken but still committed to their coffee caravan.

The following night, Willem, Mokhtar and Camilo were invited to dinner at the home of the American head of the NGO, funded by USAID, that was overseeing the entire agricultural project in Yemen. She lived close to the conference hotel, but she insisted they travel in a convoy of three vehicles, with armed contractors. The SUVs passed through the hotel gates and entered her building through another gate, past another pair of armed guards. They'd driven one block.

Before dinner, she served whiskey to Willem and Camilo and tea to Mokhtar, and they talked about the future of coffee in Yemen, and of Yemen itself. She wasn't optimistic. She'd been in Afghanistan, she said, and the situation in Yemen was far worse. It wasn't just the Houthis, she said. For Americans, the Houthis were more of a known quantity, and despite their "Death to America" slogans, their behavior so far had been, for a rebel army slowly encircling the city, more or less civilized. For any Americans, or any Westerners, really, the concern was al-Qaeda.

She told them not to leave their hotel again, and traveling anywhere outside the city was out of the question. They couldn't get a permit to do so anyway, and she couldn't help them get one.

The coffee caravan was dead. Willem and Camilo were in Yemen as guests of USAID, and the United States couldn't be responsible for their safety anymore. They had to leave.

The next day, they flew from Sana'a, headed for Ethiopia.

*   *   *

Mokhtar lay on the floor of his aunt and uncle's house, staring at the wall. Willem was gone, and he'd be booking his own flight home soon. It was all over before it began. He'd go back to California. He had to retake the Q test anyway. Maybe finish college. And there was always law school. But he needed money for all that. He thought of the Infinity. He could sleep on his parents' floor on Treasure Island and go back to being a doorman. Save money for three or four years. He'd have his undergraduate degree by the time he was, what, thirty? Night closed around him. When the 4:00 a.m. call to prayer came, he hadn't slept.

CHAPTER XXI

A DREAM IN DIFFERENT CLOTHING

IN THE MORNING, HIS young cousins woke and ate and left for school, and Mokhtar had no place to be. He had no meetings and no plan. The coffee workshop was over and he was alone. He knew nothing, so could do nothing. He didn't know much about varietals and cultivation, soil types or irrigation. He had no money and his heroes were gone.

He wandered Sana'a that day, feeling trampled upon but then again free of the burden of dreams. He had had a dream, and dreams are heavy things, requiring constant care and pruning. Now his dream was gone, and he walked the streets like a man without anything to lose. He could do anything. He could do nothing. He could even stay here, in Yemen. He walked by the University of Sana'a and for no reason he went inside, meandering through the dark ancient hallways until he saw a notice for an agricultural festival happening the next day. Everything grown in Yemen would be represented—bananas, mangos, figs, honey, coffee.

He expected little from it, but he planned to go. He had nowhere else to be. He went back to Mohamed and Kenza's house and again had a sleepless night. In the bleakest hour, though, he thought of a fruit tree he'd known as a kid. It was in the middle of the Tenderloin. There were few trees in the neighborhood—maybe there were none—none but this one. It was on Ellis Street, just a block from Glide Memorial Church, where the homeless and the city's most vulnerable lined up for food and shelter. It was a lemon tree. As a kid he'd discovered it, an actual lemon tree in the Tenderloin. At first he thought it was fake—the fruit looked too pristine, too yellow, its skin too smooth. But then he took a lemon down from the tree and smelled it; it was real. He brought it home and cut into it; it was succulent and alive.

On Mohamed and Kenza's floor, he fell back asleep thinking of that lemon and that tree and having a faint idea why.

When he arrived at the festival the next day, he almost laughed. The USAID gathering had been tiny by comparison. The University of Sana'a Agricultural Festival was outdoors and was enormous and, most crucially, it was homegrown, encompassing anyone in Yemen cultivating anything. There were almond growers, honey and guava farmers, wheat producers, purveyors of agricultural equipment and pesticides. All the coffee people were there, too. Mokhtar's pride swelled, remembering how fertile Yemen was.

Dressed as Rupert, Mokhtar went table to table, unsure how to present himself. Was he with USAID? Not really. The Coffee Quality Institute? No. Was he a student, as his grandfather had recommended? *Don't tell them you're a buyer,* Hamood had said. Willem had said the same thing. *Make no promises.*

One coffee co-op showed him their beans, which were cracked and

141

of wildly uneven quality. He couldn't help pointing this out to them. And then he couldn't help showing them photos, on his phone, of what proper cherries looked like, all ruby red, none of them green. He showed them a photo of an Ethiopian drying bed full of red cherries. The people from the co-op had never seen anything like it.

Mokhtar left the festival with a pocketful of business cards and phone numbers. In particular, a handful of people had stuck in his mind. There was Loof Nasab, a seasoned NGO worker and botanist. And there was Yusuf Hamady, president of the al-Amal Cooperative in Haymah. That night, lying on Mohamed and Kenza's floor, listening to the sounds of Sana'a at night, he had the thought that he could do his own coffee caravan. Without Willem, he would have to do all the talking. He couldn't follow, couldn't learn while quietly observing. He would have to pose as some kind of eminence, someone worthy of the farmers' time.

The next day he called every number he'd gathered, leaving messages and trying to set up appointments. He reached Yusuf Hamady. Mokhtar asked if he'd be able to show him around the farm the next day. Yusuf agreed, so Mokhtar called Loof Nasab. Would he be willing to come along on a visit to Haymah? Loof said he would. Loof, Mokhtar thought, could be a sort of Yemeni Willem, a mentor, an expert. He knew the regions and knew coffee. Beyond that, he knew very little about Loof, and tried not to worry about whether Loof would remain a helper or become an impediment. The words of his grandfather were foremost in his mind: *Don't trust anyone. Don't partner with anyone. Keep a low profile.*

Mokhtar woke at dawn. He and Yusuf planned to meet Loof at six, at the roundabout under the Panasonic building in Sana'a. Mokhtar got up, washed, tucked his bedroll into the corner of Mohamed and Kenza's living room, and went about the careful process of establishing his look. The look was important, the accessories were crucial. He had to compensate for his youth with the accoutrements of an established man.

First, the watch. Any man of means in Yemen had an impressive watch. Mokhtar's was Swiss made, silver and sturdy. Not so expensive that it would inspire jealousy or theft, but it was the watch of an executive, a world traveler.

Next, the glasses. He'd only recently realized he was nearsighted. One day a few years back, he'd tried on a friend's glasses as a joke, and suddenly the world was in high-def. He was on Treasure Island, getting off the bus at night, and when he put the glasses on he saw the crisp outline of the city, the stars, the carved lines of every wave in the Bay. He'd bought hexagonal frames, wire rimmed, made in 1941. They wrapped around the back of the ear, giving him the air of a scholar not unfamiliar with adventure.

Next, the notebook. In Oakland, he and Justin had gone to a store that sold handmade leather goods; Mokhtar wanted a durable but antiquarian-looking notebook that he could take out at key moments, jotting down crucial details from his farm visits. They picked an elaborate notebook whose pages were kept closed by a long leather strap. Never mind that he wouldn't, in the end, use the notebook or the pen much. It was far easier to type notes into his phone, where he could collate and e-mail them.

The most critical part of the uniform, though, was the ring. The ring had its roots in Yemeni history, in coffee, in the Arab Spring that had led to the downfall of President Saleh. Mokhtar had gotten it a few years earlier and Tawakkol Karman, then the youngest-ever winner of the Nobel Prize, and the first Arab woman, and first Yemeni, was instrumental in putting it on Mokhtar's finger. Karman, who won the Nobel for her work organizing Yemenis during the Arab Spring, came to San Francisco in 2011 to speak at the Boalt School of Law at UC Berkeley. Mokhtar was assigned to be her translator, and at a reception after the event, Mokhtar met one of the organizers, Mohamed Alemeri. Alemeri was wearing a ring with intricate carvings in silver, a carnelian stone inside.

"I know where that ring comes from," Mokhtar said. There was a neighborhood in Sana'a, Jawhash, where silversmiths, historically most of them Jewish, had been making these rings for centuries. When he told Alemeri this, he was astounded. In the Arab way—make note of something, compliment anything, and it will be offered to you—Alemeri insisted Mokhtar have the ring. "I can't," Mokhtar said. They went back and forth until Tawakkol Karman intervened. *"Take the ring,"* she said. So he took it and knew it would come in handy in Yemen; he could point to the rubylike carnelian stone as the ideal color for a coffee cherry.

He packed a knife, too. Not a Yemeni *jambiyah,* the traditional dagger worn in the belt, over the stomach, but an American-made knife, twelve inches long, that he wore in a leather side holster he borrowed from Hamood. Looking something like a cross between Indiana Jones and a graduate student of agriculture, Mokhtar was ready.

* * *

He left Kenza and Mohamed's apartment at 6:00 a.m. and went to the roundabout that he and Yusuf had agreed upon, the one just outside Sana'a, under the Panasonic building. When Mokhtar arrived, he saw Loof, and looked around, and immediately saw Yusuf, head of the al-Amal Cooperative. He waved to him, and then Mokhtar saw Ali Mohamed, the head of marketing and sales from a different cooperative. *No,* Mokhtar thought. *No, no, no.*

Had he double-booked? He knew he had, but didn't want to believe it just yet. Mokhtar talked to Yusuf. Yusuf told him they'd agreed on this day, this time, this location. But why was Ali Mohamed there? In the flurry of calls he'd made the day before, he must have booked Ali Mohamed, too. Mokhtar went to him, feeling his stomach in his shoes. Ali Mohamed insisted that he and Mokhtar had agreed upon this day, this time, this roundabout.

He walked back to Yusuf and asked about the possibility of visiting both farms on the same day. Maybe he could go to Yusuf's co-op first, and then to Ali Mohamed's?

"They're hours away from each other," Yusuf said. "And to do this right, you really have to spend the day with us."

Mokhtar cursed himself. What kind of moron double-books his first visit to Yemeni coffee farms? He'd immediately instilled doubt in the minds of the first two cooperatives he intended to work with. But for now he had to choose.

He chose Yusuf. He'd met Yusuf at the University of Sana'a, and he seemed the most conscientious.

Mokhtar went to Ali Mohamed and apologized, telling him that they would have to do it another time. Anytime, he said. I'll make it up to you. There'd been a mistake, one of those things.

Ali Mohamed pulled away, and Mokhtar knew he'd woken before dawn to be there at 6:00 a.m., and now he had a two-hour drive ahead of him, and all of this was for nothing.

"Sorry!" he yelled again as the truck sped off.

Mokhtar and Loof went back to Yusuf and got in the truck. Inside they met Mohamed Basel, Yusuf's right-hand man. Yusuf, tall and thin and wearing wire-rimmed glasses, was president of the al-Amal Cooperative, and Mohamed, shorter, sardonic and with a cheek full of qat, was in charge of sales and marketing.

They talked idly about Sana'a and the roads and their meeting at the university. Yusuf had grown up in the village they were going to, and had been educated at the University of Sana'a and later at the Yemeni air force academy. He'd flown jets for the Yemeni air force, but now had returned to run the collective. He was a deeply sincere and sensitive-looking man, looking less like an ex–fighter pilot and more like an assistant professor of classical poetry.

They left the capital, and the city gave way to the low-slung outer towns of Sana'a, and the highway spread out in front of them, the occasional gas station or mini-mall sitting on either side. Soon the road thinned from four lanes to two and wound in narrow bands, the truck holding tightly to the pavement as it rose and fell through steep mountain passes. The drop-off was a hundred feet, then a thousand. The architecture shed centuries and the evidence of a central government or control fell steadily away.

Most of the men they passed, old and young and teenaged, carried automatic rifles. They were in tribal territory, and as Mokhtar was getting used to this—he'd never in Sana'a or Ibb seen so many men so heavily armed—they rounded a tight hillside and found themselves surrounded. There were twelve armed men blocking the road.

Hamid stopped the truck. The men were agitated, pointing AK-47s at every window. *Who are you? What are you doing here?* they yelled. On the hillside another twenty men stood armed with rocket-propelled grenade launchers and German-made G3 assault rifles.

"What's happening?" Mokhtar asked.

"Don't worry. Tribal feud," Loof said.

The men demanded identification from everyone in the truck. Mokhtar passed his ID through the window.

Mokhtar's own tribe had been involved in a feud about ten years before. He'd been in the United States at the time, but it was news all over Yemen and among the Yemeni diaspora. Apparently a young man of his tribe, the al-Khanshali, had been in Sana'a, driving a new and prized Land Cruiser. He'd parked it one night and in the morning found it gone, stolen. Word made its way around the neighborhood about who was responsible for the theft, and the thief's own tribe, the al-Akwa, made it known that this theft would not be countenanced. Whoever the thief was, he didn't know whom he'd stolen from and didn't know the strength of the tribe. The leaders of each tribe met and made peace. In a show of contrition and respect, the al-Akwas led a procession that included not just the stolen vehicle, but, as compensatory gifts, another car and a truckload of cows. Their leader recited a poem of apology and respect, and the leader of the al-Khanshalis recited a poem accepting the gift.

Now Loof explained this was the same kind of thing—a tribal dispute that would be worked out peacefully, despite the appearance of imminent violence. Loof wasn't concerned, and Yusuf and Hamid weren't either. But a rogue armed checkpoint on the way to Haymah—this was new, a manifestation of the power vacuum, one that would benefit Saleh. The feeling among Mokhtar's family was that Saleh was trying to destabilize the country to better make his case that Yemen needed him.

As the tribesmen examined their IDs, Yusuf explained to Mokhtar that the men were looking for members of an opposing tribe. There had been a murder, and they were looking for retribution. If any of their last names belonged to this opposing tribe, the consequences would be grave.

"Don't worry. We're not involved," Yusuf said.

A few minutes later their IDs were returned, and Hamid drove again, everyone in the truck acting as if nothing much had just happened.

They drove into Haymah and, at a gas station called Abu Askr, they took a quick right—Mokhtar hadn't even seen a road there—and descended into a valley. The dirt road was rough, full of holes and covered in rubble. Mokhtar's head hit the truck roof, and he laughed it off, thinking it couldn't possibly happen again, and then it did. A dozen more times. He had to position himself with one hand on the ceiling and another on the frame so his head wouldn't hit the window. The driver didn't seem to have a consistent strategy. Sometimes he would speed up, as if wanting to simply plow through the rough road, to get it over with, and then he'd slow down, traipsing, camel-like, over the bumps and potholes.

Mokhtar was queasy, and the heat was severe, and he was ready to stop, desperate to stop, even though they'd only been in the truck two hours. But he had to maintain a façade of the routine. He was some vaguely defined representative of the Coffee Quality Institute, or USAID, or both, and presumably he'd been on thousands of miles of roads like this. Around every bend the scenery was glorious, with jagged slate-gray mountains striped with impossible terraced gardens. The architecture was simple, adobe of tan and white, the buildings sturdily built and well kept, but often perched atop seemingly unreachable peaks and ridges. Below the villages the hillsides were green with what Mokhtar assumed was coffee.

"That's mostly qat," Yusuf said.

Ten years ago, all this would have been about 85 percent coffee, he explained. Twenty years earlier, it would have been all coffee. But every year qat took more of the land.

They passed through small villages, and each time they did, the truck had to slow to walking speed, and the locals came out, wanting to know what was happening, who these men were.

Because Mokhtar and Loof were dressed formally, and had a city look about them, the villagers who approached the truck assumed they were from the United Nations or USAID or some other international entity.

They drove through a half-dozen villages before they arrived at Bait Alam, where Hamid stopped the truck. It was midmorning and already about ninety degrees.

Mokhtar got out, squinted into the sun, and found that a few dozen villagers had surrounded their truck. And now they were breaking into song.

"Peace be upon you, honored guest," they sang, "Welcome to the land of Bait Alam, where our rivers overflow and our fruits have ripened for you! The tribe of Al-Hamdan welcomes all who cross their land in peace!" This was a *zamil,* a traditional song of welcome specific to Yemeni villages—each one different and often customized for specific guests and occasions. Mokhtar smiled and thanked them, and when they were finished, the men of the village formed a line. An elder was counting them, head by head.

"What's happening?" Mokhtar asked Hamid.

"They're having a lottery for who gets to host you," Hamid said.

"Come," Yusuf said, and took Mokhtar's hand. He followed Yusuf up a steep slope into the hills. A few hundred stone steps brought them into the terraces where the coffee plants were, and for the first time Mokhtar was on a real coffee farm. He touched the leaves. He smelled the leaves. He tried to look professorial and maybe even concerned about some defect he'd found. *This is where it all started,* he thought. His euphoria lasted a minute or so.

"That's not coffee," Yusuf said.

Mokhtar had been carefully examining an olive tree.

"I know that," Mokhtar said, attempting to recover. "But the vegetation around the coffee plants affects their health."

He had made this up on the spot, and only later discovered that it was true. Yusuf nodded respectfully and they walked on.

"*Here* are the coffee plants," Yusuf said.

Now Mokhtar touched the leaves and saw a constellation of red and green cherries under the leaf cluster. The hillside was covered by an undulating grid of bright green *Coffea arabica* plants, thriving on what looked to be an arid mountainside. There was the smell of jasmine, the faint hushing of the breeze passing through the dense foliage.

"So what do you think?" Yusuf asked.

"Good?" Mokhtar said.

He didn't know what Yusuf was looking for. They walked on and soon were joined by a growing throng of farmers and pickers who asked a series of questions:

"The leaves are being eaten by burrow worm. What should we do?"

"What about pesticides?"

"What do you think of the soil here?"

"What's this white stuff ringing the trunk of this one?"

Mokhtar had no idea. He was no agronomist. This was the first coffee farm he'd ever been to. Not that he could tell anyone this. But Loof was an agronomist, and Loof stepped in.

"That's sodium," he said about the white rings. "This plant is getting too much salt." He began answering the questions, touching the leaves, squatting down, inspecting the soil. He had answers to all their queries, and Mokhtar activated his Tenderloin memorization brain, processing all that Loof said, ready to regurgitate it later. Loof went into the merits of pruning, explaining that every tree was like a family, each bough a child, and that a plant could only support so many healthy boughs—that any boughs that weren't viable needed to be pruned. He pointed out the different varietals, names of which most of the farmers and pickers didn't know.

"This is a Tufahi," he said. "This is Dawiri. This is Udaini."

The farmers had simply been growing coffee, the generalized coffee of the second wave. Yusuf was aware that there was something out there in the wider world, something changing, some new attention being paid to regions and varietals, but his co-op didn't have enough information or access. They didn't know which varietal was which, which thrived where, how best to pick and process cherries of these varying kinds and, most of all, who would pay for it.

Mokhtar was careful. He had a sense that somewhere down the line he could help with the supply chain, and he hoped that he might be able to connect these farmers with high-end buyers in the U.S. and Europe and Japan, but he couldn't say any of that now. His grandfather had drilled that into him: *Don't make promises unless you're sure you can deliver. And don't make them* until *you have the funds to deliver.*

For the time being, then, Mokhtar walked with Yusuf and Loof and he listened. He listened to Loof talk about how best to pick the cherries and when to pick them. He observed Loof's way of speaking, his mannerisms, and tucked them away for future use. And he tried to keep up as Yusuf and the older, and even elderly, growers and pickers moved up and down the terraces like rabbits. Mokhtar had to be pulled back from cliffs and caught after slipping on steps. The air was thin, and he had to catch his breath. His unsteadiness entertained the local men.

"Who's that?" Mokhtar asked.

Mokhtar noticed one man sitting alone, under a tall and extraordinarily healthy coffee tree.

"That's Malik," Yusuf said. "Our best farmer."

He sat cross-legged in the shade, looking supremely content.

"He does that a lot," Yusuf said. "He's always out here. When he's not picking, he's sitting out here among the trees."

Malik wore a gray pillbox-shaped hat called a *kufi*, adorned with complex embroidery. Mokhtar was intrigued by the man's aura and attire, and took a few pictures, noticing that the man had placed his day's pickings on a towel by his feet. He had about five hundred cherries there, all ruby red.

"That's why he's our best farmer," Yusuf said. Apparently Malik did most of the picking himself, with the help of his wife and a few family members. For them it was not a job or hobby or something delegated to careless hourly workers. It was a calling, something they enjoyed and in which they took a spiritual kind of pride.

In the cooperative, Malik owned perhaps four hundred coffee plants, which stood next to those owned by another farmer. All of the farmers picked their own coffee, which was then mixed together in a mass of cherries, red and green, ripe and rotten. These were then sold to a broker, who usually exploited the financial disadvantage of the farmers. There was no separation of lots or varietals. It was all one heap, sold for whatever price the brokers offered.

Mokhtar approached Malik as he sat under his tree. The man did not stand up or act in any way deferential. In fact, he didn't seem the least bit surprised or excited to meet this man from Sana'a and America. But when Mokhtar asked if he could take a sample of his cherries back to Sana'a, Malik respectfully agreed.

"We can give you a big bag later on," Yusuf said.

"I want these," Mokhtar said. "I want this man's cherries, and I want them separate from all others."

*   *   *

For lunch Yusuf brought them to his home. Somehow he'd won the lottery conducted when Mokhtar first arrived. Yusuf's house was a traditional Yemeni rural dwelling: the first floor was open and dark, intended for livestock. The second floor, and every floor above, was dedicated to a branch of the family. Each of the home's seven floors was called a house, and each was given to a different family, or families, everyone related. In Yusuf's home there were four generations.

Mokhtar and Loof were treated to a feast, though one belying the relative poverty of the village. The chicken was lean and gamy. The rice was plentiful, and there was bread, and *sahawqah,* a salsa-like dish made from the hot peppers the region was known for.

With the meal, served with great ceremony, Mokhtar and Loof were given coffee, which Yusuf noted was made from the coffee trees they'd just walked among. But it wasn't coffee, not the kind of coffee consumed anywhere else in the world. This was a beverage made from the dried husk of the coffee cherry. It was what Mokhtar and most non-Yemenis called *qishr*—a sweet kind of tea, caramel colored but with a certain sweetness, a ghost of the cherry, in its center. It was delicious, but it wasn't coffee. Mokhtar didn't know how to tell Yusuf this, and knew it didn't bode well: if the president of the cooperative didn't know the difference between coffee and tea, then there were more problems than he'd anticipated.

After lunch they sat for qat, and were joined by dozens of men, most of them farmers, many of them other villagers who were interested in the visiting Yemeni American and his Sana'a friend. Everyone was intrigued by Mokhtar, all but one man. Mokhtar had seen him here and there throughout the day. He was a stout man wearing the coat of a former soldier and what looked like a Russian hat, with furry earflaps folded upward. He had an AK-47 slung over his shoulder, and two grenades fastened to his chest. Mokhtar had caught his skeptical eyes and asked Yusuf about him.

"That's the General," Yusuf said. "Don't mind him. Hard to please." He had been a general in the Yemeni army and upon retiring, had bought land in the Haymah Valley, where he planted both qat and coffee; he was one of the largest landowners in the region. The General glared at Mokhtar throughout the meal.

Loof wouldn't chew qat. He never did. Mokhtar thought it best to indulge, so as not to offend their hosts, but Loof wouldn't budge. And besides, this was a different style of qat enjoyment. In the city the qat would be pruned neatly, prepared and presented carefully. Here it was dumped in the middle of the floor like kindling. But Mokhtar took leaves and filled his cheek, and they all talked idly until the qat had kicked in. A wave of mild euphoria spread through the room, and Mokhtar chose the moment to present them with the past and future.

He told the men about the birth of coffee, that it was first cultivated here, in Yemen, that it was a central part of the country's history, their birthright. Most of the men seemed surprised by this. Had they known this? He wasn't sure. He went on, explaining that the Dutch had stolen the seedlings, had planted them in Java and had given them to France, and the French had planted them in Martinique, and that the Portuguese had smuggled them from the French, had planted them in Brazil, and that now there was a seventy-billion-dollar market for coffee, that everyone seemed to be making money from the bean—everyone but the Yemenis, who had started the whole business in the first place.

Maybe it was the qat, but he had their attention. Even the General was listening, albeit with a look askance. Most of their beans were going to Saudi Arabia, Mokhtar explained. The farmers were selling them for next to nothing, and that needed to change. But first they had to improve their practices. They should harvest cherries only once they were red—and here he showed them his ring, with the carnelian stone inset. And then you need to dry the cherries on aboveground beds, so air can circulate and they dry evenly. Then you need to store them properly, in cool and dry rooms, so they don't ferment or accumulate mold. Right now, he explained, you're picking too soon, and the green and yellow and red cherries are being mixed together, and dried improperly, and shipped recklessly, and sorted carelessly if at all. The roasting was a disaster in Saudi, he said, so all along the chain the plants were being disrespected, the beans abused.

He went on, telling them about the Monk of Mokha, about how they needed to reclaim their heritage, and how if they improved their process, if they picked better, dried better, stored and shipped better, the prices might be higher, their wages might be higher.

"Will you help us?" a man asked, and Mokhtar caught the tiniest hint of expectation in the General's eye.

Me? Mokhtar thought. *Not yet at least.* He hedged. Though he wanted to, he couldn't tell them he was looking at all of this not just as a consultant, not just as some vague representative of the CQI or USAID, but as a potential buyer, a potential exporter.

After qat, he and Loof were asked to sign the village registry. Every visitor for centuries had signed it—an enormous book with yellowed pages. Mokhtar signed his name and under it he wrote, *With your sweat and blood and hard work, your coffee will be the best in the world.* It seemed like the right thing to say.

On the drive back, Yusuf was visibly excited.

"So you can help us get better prices?"

"I don't know," Mokhtar said.

"But if we improve our processes, we can get a higher rate?"

"I'm not sure," Mokhtar said. *"Can* you improve?"

They got back to Sana'a late. The city was quiet, and Mokhtar let himself into Mohamed and Kenza's house and put his coffee samples under a chair in the corner, unspooled his bedroll, and lay down.

I can't do this, he thought. *No chance.*

Somewhere during the long drive home, winding their way through hundreds of miles of two-lane roads, passing countless armed strangers, thinking at any moment that their truck would be stopped again at a checkpoint, Mokhtar's doubts overtook him. This was the first trip, and he'd already faced a hillside of vengeful tribesmen. He'd bluffed his way through one farm visit, but this was over his head. It was madness.

And there were the loan sharks. He was going to go head-to-head with loan sharks. Loan sharks he knew nothing about. His grandfather was from a powerful tribe, but was Mokhtar ready to cut a bunch of bloodthirsty moneylenders out of the process? Clearly their scruples were questionable—they were subjugating these farmers and had no respect for the quality of the coffee they were selling. What would happen if Mokhtar from San Francisco came in and cut them out of the loop?

And these farms were disastrous. Some of those beans had been stored for five years! The farmers were sitting on them like imperishable currency, as if the beans never aged. And the thing with the tea—did they know the difference between tea from the husk and coffee from the bean? And could he really improve harvesting practices such that the coffee was actually worth significantly more? And who knew if it was any good in the first place? That is, even if they picked it correctly, and dried it correctly, and processed it correctly, and did everything else correctly, who could predict if the coffee was actually any good? The coffee made from these beans might be terrible. And that would be a fact that no amount of supply-chain adjustments could remedy.

And again, he knew nothing. He'd been studying coffee for months, and learning from the best—from Blue Bottle and Willem Boot—and that had given him some knowledge about cupping and roasting, but he didn't know anything about growing these plants, or harvesting, or sorting. He knew nothing about the actual plant growing in the actual world. Loof had run circles around him.

Nope, he thought. He was a clown. He had a company name, and a logo, but he didn't know what he was doing. He'd borrowed money from Omar and now it would all be wasted. He could and should go home. He could go to college, study something. He'd lived long enough as a corner-cutter. With this he couldn't play *Fake it till you make it.*

CHAPTER XXII

POINT OF DEPARTURE

THEN AGAIN, IT HAD worked before.

The roundabout became his point of departure. Almost every day for the next three months, he went down to the Panasonic roundabout and got into a different truck, every time venturing out to a different region, determined to visit every one of the thirty-two coffee-producing areas of Yemen. Some were like Haymah—hardworking and reasonably advanced, with farmers he knew that he could work with. Other visits were disheartening. One time he drove seven hours to find that the region was home to only a handful of coffee plants—fewer than twenty. Some farmers he simply didn't trust.

And the skepticism was mutual. In most of the places he visited, the farmers treated his arrival with polite suspicion. They'd been visited by NGOs more or less continually for decades, and by USAID more regularly after 9/11. Sometimes progress was made, sometimes not. Occasionally a water catcher was built; other times that or any other project was begun but not finished. The intentions of these aid workers and organizers were fine, sometimes unassailable, but the follow-through was inconsistent. This was the backdrop of Mokhtar's arrival. He would come, well dressed and speaking his classical Arabic, with his American passport and though they wanted to believe he had answers, insight and, most important, some way to get their beans to market for a higher price, they were reluctant to believe.

Still, customs of hospitality dictated that he be treated well. So they would show him their terraces, they would feed him a midday meal, in the afternoon chew qat, and occasionally put him up for a night. Did they expect to see him again? Not entirely. Did they expect him to change their lives? No.

He knew he could improve their cultivation methods. He knew that by merely ensuring that their pickers chose only the cherries that were ripe, the color of the carnelian stone, their quality would rise dramatically. That pruning would increase the production of the plants, and if they dried their cherries on raised drying beds, the quality would rise yet more. If they bagged their beans within plastic, not burlap, they would hold in moisture and improve the taste yet again. These were basic things they could do on the farming side of it. He knew that if he got hold of a better-picked and better-processed class of cherries, he could process them more carefully than had been customary, and then, once they were beans, he could sort them far more carefully than had been done in a hundred years. He was sure his impact could be real.

The challenge was staying alive until the next harvest. In his travels this would prove difficult. In the first week, he came down with malaria. He was in Bura'a, in far-western Yemen, and he woke up with yellow eyes. He'd gotten it either in Bura'a or Haymah, but here it was, and he couldn't move. He had the shakes. His limbs felt withered and weak. His Bura'a hosts gave him pills, but he was sure he would die. They brought him to a local hospital, where he spent two feverish nights.

After recuperating in Sana'a, he went out again, and this time, in Bani Ismail, he was overtaken by a maniacal kind of diarrhea. He spent two days circling the shithole—there were no toilets in the village—feeling sure that the next thing to leave his rectum would be his liver.

A few weeks later, he was visited by what he and everyone assumed was a tapeworm. He ate all day and all night and couldn't gain weight. Someone said it was his body correcting for the loss of weight incurred during his bouts of malaria and diarrhea. He ate still more and somehow got even thinner.

You can live with it, one friend said. *Coexist.*

Try kerosene, said another friend.

Apparently this had been a longstanding practice—the drinking of kerosene to kill the parasite. Mokhtar decided to wait it out, and after another week his metabolism returned to normal. He had no idea if the tapeworm had exited on its own accord or if he'd had one in the first place. He had a few weeks to enjoy the functioning of his digestive system before he knew the once-in-a-lifetime pain of a gallstone. He spent another night in the hospital for that, and left a withered man.

He stayed in Yemen three months and was sick every four or five days. He'd been told to be careful with the drinking water, with fruit, with anything that might harbor bacteria—that he was an American and unaccustomed to whatever organisms lived in and were tolerated by Yemenis. And though he knew he should refuse certain foods

161

in certain villages—or most foods, all uncooked foods, all water, all juices, all fruit, in all villages—he couldn't. He was a guest, and he was a guest needing to be respectful, needing to emphasize his own Yemeni heritage, and not underline his foreignness or preciousness. So he ate whatever was put in front of him and hoped for the best. He got diarrhea too many times to count or care anymore. It was, in the end, a small price to pay as he benefited from the Yeminis' legendary generosity.

He continued to go out. Back in the car. Over the rutted roads and through the narrow mountain passes and into new villages, where each time he was greeted by men singing traditional *zamils,* followed by the lottery to determine who would have the honor of hosting him. For lunch and qat always they would arrange mountains of pillows and blankets at the head of the room, propping up Mokhtar like a Mongolian warlord. There was always cold soda—upon Mokhtar's arrival, the village sent its children miles on foot to get it. And after the tour of the terraces and after lunch and qat, there were gifts, always gifts. If the region was known for mangoes, Mokhtar would leave with more mangoes than he could possibly eat. If honey was made there, he'd leave with enough to fill a tub. And of course each time he left with coffee cherries, a sample of that village's best, and he'd return to Sana'a to put the samples in a corner of Mohamed and Kenza's living room, and he'd go back to sleep.

He went to Bait Aaliyah, two hours from Sana'a and over two thousand meters above sea level. Blessed by a plentiful aquifer, its farmers cultivated thirty thousand trees. He went to Bani Matar, about two hours from Sana'a and eighteen hundred meters above sea level. In Bani Ismail, he saw the most prized and expensive of all Yemeni coffee. The beans were small and almost circular, heavily dependent on rainfall. The local cooperative was amiable and organized, but they weren't sure how much coffee they produced. In a given harvest, they said, they filled two trucks a week, for about eight weeks. Any more detailed measurements, they hadn't done. In Al-Udain he saw the most visually beautiful beans in all of Yemen. But the trips were not universally successful. In fact, most were not. One day he drove seven hours to Hajjah, which his research said was a coffee-growing region. When he arrived, he found no coffee plants. A local farmer walking on the dusty road was astonished that Mokhtar had come so far for no reason.

"My grandfather's grandfather grew coffee," he said. "You're about a hundred years too late."

Every time Mokhtar returned to Sana'a, Mohamed and Kenza would note the arrival of another bag, and would note that their living room seemed to be shrinking. But they said nothing to Mokhtar, assuming that what he had told them—that he was writing a report on coffee in Yemen—was the unalloyed truth.

After a while, though, he couldn't deceive them anymore.

"I'm starting a business," he told them.

"Why didn't you tell us?" they asked.

The answer was complicated, he said. Hamood had insisted he keep his plans close to the vest. But then there was the matter of how Yemenis saw the domestic coffee business. It wasn't taken seriously. Saying you were in the coffee business was like saying you sold lollipops. No one made a living selling coffee.

"But this can work," he told them.

He told them where he'd been and showed them his photos. They were awed. They'd never seen those parts of Yemen. They'd been to Ibb, sure, but not Haymah, not Bura'a, not Hajjah, not Bani Hammad.

"Why didn't you take us with you?" they said.

Sometimes he took their son, Nurideen. At eighteen, he was the oldest of their six children still living at home. He was newly out of school and without prospects. Thinking about college in America, he'd applied for a visa at the U.S. embassy and had been turned down in a comically dismissive way.

"Who's sponsoring you?" he was asked.

"My brother Akram," he said. "He works as a janitor at the Contemporary Jewish Museum in San Francisco."

"San Francisco? That's the most expensive city in the world!" the agent said, and rejected the application.

After that, Nurideen applied for and received a visa to live and work in South Korea. He flew to Seoul, but when he landed, they wouldn't accept him. He had all his paperwork in order, but they turned him around and put him back on a plane. He flew to Malaysia instead, a country historically hospitable to Yemenis, and worked for a time in a restaurant where he was underpaid and abused.

Now Mokhtar needed help. He needed someone to come with him to provinces, to help catalog the samples and keep track of the farmers and harvests. Nurideen became Mokhtar's first employee, with the matter of a proper salary, of course, deferred.

* * *

Mokhtar continued to go into tribal areas, hours or days from Sana'a, and every time he packed his dagger, and a SIG Sauer pistol. His driver had a semiautomatic rifle. When he was in more troubled or unknown districts, he brought along another man who carried an AK-47 and a grenade. None of this was unusual. There were twenty-five million people in Yemen and at least thirteen million guns—after the United States, it was, per capita the world's most armed nation. Men wore AKs walking down the street. They brought them to weddings.

When Mokhtar was young, Hamood had given him a pistol, a Colt .45, and Mokhtar still had it. Eventually he'd bought an old AK-47 and occasionally borrowed Hamood's 1983 Krinkov. He liked to have them in the truck in case they got caught in a tribal dispute, or if someone tried to get at the cash Mokhtar strapped under his belt. Or in case they needed gasoline.

Rumor had it that those supporting the ousted Saleh bombed the petroleum pipelines. He wanted to sabotage the infrastructure, to convince the Yemeni people that things were better when he was in charge. And so gas sometimes grew scarce, prices skyrocketed and gas lines grew long. When the lines were long, tempers went hot. Someone tried to cut the line, guns were drawn, shots were fired in the air.

Mokhtar got so used to being out in the provinces, returning to Sana'a dusty and unwashed, unshaven and dressed in a tribal way, that he forgot, momentarily, to which world he belonged. Once a week he'd be in the capital, and he'd go to the Coffee Corner Café, an upscale place frequented by wealthy and worldly Yemenis, and some Westerners, and there he'd use the wifi and write up his reports.

One morning he walked in, unshowered and ill slept, and he sat near a pair of young women wearing expensive shoes and bright hijabs, their makeup sophisticated and subtle. One of the two pulled out a laptop and began watching the *Vampire Diaries* without headphones. The whole café could hear the show, its wailing and screaming. The two young women didn't care. Soon, though, their attention focused on Mokhtar.

"Look at him," one of the women said. "He's so barbaric and backward." It took Mokhtar a second to believe she was talking about him. She spoke in English, thinking Mokhtar was a peasant who had somehow wandered into this sophisticated urban café. He was wearing his dagger and carrying his pistol, and between that and his disheveled state, he understood that she took him for a tribesman from Yemen's northern hinterlands, who were seen as rough and violent and were often derided by city dwellers.

The Houthis adhered to a branch of Shia Islam called Zaidism, which accounted for about 35 percent of the Muslims in Yemen. Before 1962, the Zaidists had controlled northern Yemen for a thousand years, and the Houthis frequently clashed with their neighbors over territory, with the Saudis to the north and the Yemeni government to the south. In Sana'a, they were considered a nuisance, uncivilized hillbillies bent on wreaking havoc.

"He's the problem," the woman continued. "Men like that are holding the whole country back."

Mokhtar had work to do, and he was tired, but the orator in him was awake and ready. "Excuse me, ma'am," he said in English. "*You're* the problem."

The woman's mouth dropped open. She looked at Mokhtar like he was an animal that had somehow learned to speak.

"You're degrading me," he continued, "while watching your adolescent show without headphones."

The women watched his mouth, as if struggling to discern whether or not he was being dubbed. They couldn't figure out how these words, in American English, were coming out of the mouth of this savage.

"You need to respect me," Mokhtar said, "and you need to respect this space, and the people here, and not make assumptions based on how I look. And actually, I think you should leave."

And they did.

Looking tribal had its advantages. The next time Mokhtar was in Haymah to talk with Yusuf and Malik about drying beds and the next harvest, he woke to a cracking of gunfire in the valley.

He packed his AK and followed the sound to a valley where he found a group of men conducting a target-shooting contest. The General was among them. The General noted Mokhtar's AK slung over his shoulder.

"You know how to use that?" he asked skeptically.

"I do," Mokhtar said.

The men were aiming at a small white rock resting on the ridge about seventy yards away. No one had hit it.

"Your turn," the General said.

Rafik and Rakan had taught Mokhtar to shoot .22s, handguns and AKs at 5 Dogs Range in Bakersfield, just down the road from Richgrove, and Rafik had educated him about the ammunition used by different firearms and the relative accuracy of each. The men shooting that morning were all using modern AKs, which were powerful and efficient, but less accurate than the AK Mokhtar carried, a pre-1974 model. For target shooting, it was the superior tool.

Mokhtar stepped up, aimed, exhaled and shot.

The rock flew off the ridge.

He stepped back, accepted the congratulations of the men and saw something like respect in the face of the General.

Knowing he probably couldn't make the shot again, and knowing the value of a well-executed mic drop, Mokhtar shouldered his gun and left.

CHAPTER XXIII

OUT OF SANA'A

MOKHTAR CHECKED THE PLANES leaving Sana'a and found one going through Qatar. Mokhtar had to get back to the United States. He needed to test the samples he'd collected—he planned to bring twenty-one lots home—and visit family and see about raising a few hundred thousand dollars so he could come back and actually buy the coffee, if any, that scored well.

He spent five days frantically finishing collecting his samples from up and down central and northern Yemen. This was during Ramadan and during the Houthi takeover of the city of 'Amran, the last northern defense before they could capture Sana'a.

Mokhtar was up until 4:00 a.m. every night milling his samples and going to sleep with coffee dust on his head. Finally, the day before the flight, he packed his clothes. He packed the samples he had at the apartment. Half of his beans were in Ibb—he was storing some with Hubayshi—but they would have to wait. Damn this country, he thought. If this were Amsterdam, he could just FedEx a box tonight, any night. He could leave and call Samir and ask Samir to send them,

169

or Mohamed, or anyone. But in Yemen anything you wanted to get out of the country in any timely way you had to take yourself.

He bought five suitcases and started filling them with twenty-two samples from twenty-one farms—every coffee from every region that had any to offer. What else? He had to get honey. His parents wanted Yemeni honey. They also wanted Yemeni almonds and raisins, so he made sure to go to the old quarter of Sana'a. He called his cousin Nurideen and asked him to help. Nurideen was awake—everyone was awake; it was Ramadan—and together they hired a taxi and spent the small hours rushing around the city gathering everything Mokhtar needed. A dozen gifts for everyone he could think of back in California—postcards, frankincense, aloes, prayer beads, silver rings with carnelian-stone settings, handmade Kashmiri shawls.

When Mokhtar's extended family learned he was flying back to the U.S., a cousin asked him to look after their six-year-old daughter, Dena, who was flying on the same plane to California. Later, Mokhtar would have a hard time explaining how an arrangement like this was made so casually—the transport of a cousin's daughter, a girl he'd never met, across continents and seas. But getting in and out of Yemen was so difficult, and it was rare when someone like Mokhtar was leaving and could escort a girl like Dena, who was going to see her family in Modesto.

So while Dena packed, Mokhtar and Nurideen continued racing around the sleepless city, feeling very much alive and laughing about it all as the sun was coming up. Then they turned a corner, onto one of Sana'a's busiest thoroughfares, and drove directly into an active firefight.

Machine-gun fire cracked open the morning. Mokhtar looked up to see the ends of AKs peeking out over the rooftops of buildings on opposite sides of the street. Their driver should have backtracked quickly, but he hadn't moved.

"Reverse, reverse!" Mokhtar yelled.

"There's no reverse!" the driver yelled. "Get out and push!"

Mokhtar and Nurideen got out and pushed the taxi backward. They laughed. They couldn't help it.

"Been nice knowing you," Mokhtar said. He figured the odds of survival were about sixty–forty.

As they pushed the taxi, Mokhtar noticed a propane tank attached to the trunk. This was common in Yemen, given the gasoline shortages—drivers rigged their engines to run on propane.

Mokhtar and Nuri laughed harder. They were pushing a taxi with an exposed propane tank while machine-gun fire rattled over their heads. They couldn't run away. All their coffee was in the taxi.

An hour later, Mokhtar was at the airport, sitting in the waiting area, thinking about all this, the time an hour earlier when he almost died. He looked over and remembered that he had a six-year-old girl with him, and that this girl hadn't spoken to him yet. Her mother had kissed her forehead and told her to be good and not to make any trouble on the way to America.

She was a beautiful child, with huge dark brown eyes and a tangle of black hair. She was wearing a Hello Kitty shirt and was carrying a SpongeBob backpack and seemed strangely uninterested in Mokhtar, the man taking her on this journey, a trip that would likely keep them together for about twenty-six hours, through Qatar, over the Atlan-

171

tic to Philadelphia, and on to San Francisco. Dena was unfazed by the whole enterprise—leaving Yemen, leaving Yemen with Mokhtar, whom she barely knew—traveling across deserts and oceans.

"You're not planning to talk to me?" he said.

She looked at him, said nothing, and looked away. She said nothing all the way to Qatar. There were movies on the plane and Mokhtar badly needed sleep. He woke up as they were landing at Qatar's Hamad International Airport. They had a ten-hour layover, so Mokhtar bought her lunch, and she ate it contentedly, and eventually Dena fell asleep on his shoulder as they waited for the flight to Philadelphia. She slept through much of that leg, and when she wasn't asleep, she ate the plastic-wrapped airline food and watched seven hours of cartoons.

When they landed in Philadelphia, Mokhtar held Dena's hand—she was so sleepy she allowed it—and they approached customs. There were two lines, one with a younger man at the booth, the other staffed by an older man, and though neither line was very long, one was faster than the other, so Mokhtar chose that one, and soon found himself greeting the younger man. Later he wondered if this had been an error. It was an impossible game that he and many Arab Americans had been trying to play for years now—was it the younger people in positions of authority who were more likely to be enlightened, understanding, raised in a more diverse and connected world? Or were the older ones, who had seen more of the world's travelers come through U.S. airports, more understanding?

"Hi!" Mokhtar said in his most American way, to show he had no

172

accent, that he was American raised. It made no difference. Within two minutes the young official put Mokhtar's passport in a red envelope.

"Step over here," the official said. "Don't worry. You're not in trouble."

Mokhtar and Dena were led into a side room, and when the door opened, he saw a small sea of Arab faces. Strangely enough, though he'd flown plenty inside the United States and in and out of Yemen, he'd never been selected for any secondary screening, no additional interviews, nothing. The novelty, seeing this room firsthand after hearing about it for so many years, activated his sense of the absurd.

"Salaam alaikum!" he said loudly, giving them a big semicircle wave. Some laughed. Most returned the greeting: *"Wa alaikum assalam."* Others seemed too worried or tired or numb. Some had been waiting five and six hours. This was the room where time ceased to have meaning, and some of the men and women in the room seemed far beyond their patience.

He and Dena sat for a spell, until Mokhtar was approached by a soft-faced official. His nametag said his name was Joel.

"Hi Mokhtar," he said. "Can I call you Mo?"

"No," Mokhtar said, and then couldn't help himself. "Can I call you Jo?" Joel smiled indulgently, and Mokhtar smiled, trying to communicate that the screening process was inherently flawed and racist, but he would, for now, maintain a sense of humor.

Joel seemed apologetic. He said it was just a formality, all this, as they left the room and went down a hallway to a baggage area, where Mokhtar had to claim his suitcases. When Joel saw how many suitcases Mokhtar had checked, he was intrigued, but continued to smile, saying it was not a problem, that this was just a formality, a normal process.

With the help of another official, the suitcases were taken to a steel table and opened, exposing all his beans in so many plastic bags. Mokhtar knew this would be far too interesting to Joel, and to customs in general. He knew he'd miss his connecting flight. He thought of lawyers he could call on the East Coast.

"So what do you do?" Joel asked.

Trying to suppress his frustration, Mokhtar said he worked in coffee, he was an importer, he was helping to improve conditions for Yemeni farmers—and by the way he was doing much of this work with the cooperation of USAID, that he was helping the U.S. government. "*My* government," he said, his voice rising. "I'm helping us look good over there!"

Now Joel was interested, but the tone of the conversation seemed to have changed. He asked Mokhtar about coffee, about which coffees were the best, did Mokhtar think dark roast or light roast was superior, and which was preferred by experts? Mokhtar calmed a bit, and talked with as much levity as he could about varietals, and different roasts, and the effect of elevation on coffee fruits, the relative advantages of Yemeni coffee, how Joel should ask for it at the next café he went to, that soon enough Joel would be able to order coffee imported by Mokhtar himself. And because all this coffee talk seemed to be going so well, Mokhtar allowed himself to believe that he and Dena would soon be permitted to zip up all the suitcases and be on their way.

But first, Joel said, there was an agricultural screening. All the suitcases were zipped up again, and they pushed them down another hallway and into another room, where they were put on steel tables again.

A uniformed woman told Mokhtar he couldn't import all these green samples without documentation and permits, and he lost track of all the things she was saying, because he hadn't thought of any of this while he'd been packing the beans. He'd survived what might have been a racist interview, but now was confronted by a very legitimate question from this reasonable woman about the legality of importing six suitcases of green beans, which may or may not be bringing along an invasive species or unknown bacteria.

The woman seemed unsure what category coffee fell into, though. It wasn't a live plant, after all. These were beans. And while they were talking, she cycled through the usual order of conversation, just as Joel had. You're from Yemen? And there's coffee in Yemen? Fertile areas? Things really grow there? I love coffee, she said. Is this good coffee? Can I get it at the grocery store? Does Starbucks sell Yemeni coffee?

Incredibly, after ten minutes with the agricultural inspector, Mokhtar was rezipping the suitcases and was allowed to proceed. It was either the power of coffee or the power of his charm, but he was on his way, holding tight to Dena's hand, and feeling very good about all this, and feeling almost sure he would make their connecting flight.

Then Joel led him to the back of the screening line.

"You have to go through security again."

He put his carry-ons through, put Dena's SpongeBob backpack on the conveyor, thinking this was the last stage, and not so onerous, really. But after the X-rays and scanners, he and Dena were pulled aside again, and while TSA swabbed his carry-on bag for explosive dust, he turned to see Dena being frisked.

They were cleared, and they walked down the hallway, looking for the connecting flight, but because they'd come through the agricultural screening, they'd ended up in a strange part of the airport, far from their gate. The TSA staff had directed him where to go, left and right and left again, but now he was lost and outside the security checkpoint again. The only way to get to the gates would be another screening.

So they went through another screening. After the screening, while rushing to the gate, with a few minutes to make the flight, Mokhtar was met by a random TSA official.

"Can I ask you some questions?"

The questions were about Mokhtar's trip to Yemen, his work, his residence in the United States. The questions went on for ten minutes, long enough that Mokhtar and Dena missed their flight.

They'd already been at the Philadelphia airport four hours. The next flight was six hours away. Mokhtar went to the airline desk and was greeted by an African American woman, who apologized for his trouble and issued him and Dena new tickets, which had the two of them separated.

Mokhtar asked for seats together, and the clerk said that was possible, but that he would have to pay extra.

He paid, and she gave him his new tickets, which bore the code that indicated they'd been singled out for extra screenings.

"You know what?" he said. "You work in a racist institution.

You should know about these things. I've been through four hours of screenings and I missed the flight. That's why I'm here getting a new flight. And you're putting me through another screening because I'm brown."

Mokhtar was on a roll, his voice rising. People around him were listening. He went on—*Are you in a union? You work for a racist organization. This is a racist system*—until the African American agent, and the white male agent next to her, began apologizing and until the white man came around and leaned down to Dena.

"You want a sticker?" the man asked.

"We don't want your stickers!" Mokhtar said. "We want dignity."

Everyone at the gate was listening. A few people clapped. He fumed until the flight boarded, and when he gave his ticket to the agent and the ticket went through the machine, another signal alerted the agent to pull him aside. She looked around and said, "Just go." He took Dena's hand and flew.

CHAPTER XXIV

THIS ONE'S INTERESTING

"YOU LOST WEIGHT." THAT was the first thing everyone at Boot said. Mokhtar had lost twenty-five pounds. "You lost your ass," they noted.

Mokhtar and Stephen spent ten hours cleaning the Yemeni samples. They were dirty, unsorted, full of broken beans and defects. Once they had something to work with, Stephen carefully roasted all twenty-one varietals, and then Mokhtar and Stephen and Willem held a very official cupping session, to determine whether or not Mokhtar, and by extension Yemen, had any hope in the world of specialty coffee.

The first batches were dismal.

"DOA," said Willem about one after another.

They were at Boot Coffee in Mill Valley, and Mokhtar was in the moment, hoping for the best with these beans, picturing the farmers he'd met, the General and Malik and Yusuf, whose hopes rose and fell on whether or not their coffee could garner a higher price, a premium price.

"DOA," Willem said about another varietal.

That day they cupped ten varietals, and Willem's verdict for all ten of them had been the same—not viable. The samples were dirty, earthy, muddy, old and overfermented. Unclean and without merit.

Mokhtar hadn't expected anything extraordinary. He figured if he had a few samples that scored in the 80s, he could work with that. He could bring those up to 90 over the course of seasons or years. But so far none of his coffees had cupped over 70. Coffee this bad would not warrant going back to Yemen. There would be no point.

The next day, Stephen carefully roasted the remaining eleven samples, doing all he could to increase the chances of the coffee clawing its way to respectability. "DOA," Willem said again.

Five of the last ten had no value whatsoever. They hadn't justified the expense of bringing them back from Yemen. Mokhtar didn't know if there was any point in continuing. He didn't think he could hear the letters DOA again.

Then Willem made a sound. A surprised sound.

"This one's interesting," he said.

A few days later Mokhtar was standing outside Royal Grounds Coffee, a major regional roaster and importer based in Northern California, and he was crying. Willem, Jodi and other Q graders had scored three of Mokhtar's samples in the 90s. Two were from Haymah, one from Ibb. One was Malik's.

Mokhtar had taken the samples to Royal Grounds, and based on the scores, they'd said they would buy eighteen tons. Mokhtar didn't have eighteen tons of anything, but in theory he could get it. If he could pay for it.

Again he went to Omar. He told him about the scores, about

the promise of significant orders from major roasters and retailers. Omar assembled a small group of investors, all of them Arab Americans who had done well in the tech sector. Together they arranged to loan Mokhtar the funds necessary—about three hundred thousand dollars—to buy a containers' worth of coffee. There were contingencies, of course. The coffee had to be of the highest quality, and the coffee had to somehow get shipped out of a highly unstable country. And they could not release the money to Mokhtar until he proved the value of the coffee and proved he could get it out.

Mokhtar agreed. He didn't know any better and had no other options. And he believed he could do it, even though the last time he had significant amounts of cash he'd put it in a satchel and lost it in a parking lot.

His parents were proud of his work in Yemen, but didn't want him to go back. He'd returned twenty-five pounds lighter than when he'd left, his skin pale and his eyes sunken. The malaria and the tapeworm—whatever that had been—had wrecked his body. They worried about his health, but more so about the fact that the Houthis had overtaken Sana'a a few weeks earlier, on September 21, and it was possible the country would devolve into civil war.

But Mokhtar bought his ticket. The next harvest was coming soon and now he could actually buy coffee. He had to buy coffee—his investors expected it, and he had Royal Grounds waiting for it. Now it was just a matter of finishing his Q-grading test, so he could return to Yemen as the first Arab Q grader in history. Easy.

Not easy. He took the test again. Again it was Jodi administering it, and it was just as difficult as before, but because he'd come so close before, and because, he thought, he was now in touch with the origins of coffee, and was on a mission propelled by fate and God, he passed.

In September of 2014, Mokhtar became the world's first Arab Q grader of arabica coffee, and in October he returned to Yemen, to Haymah. He wanted to visit Malik, the man he'd first seen under the coffee tree. It was his coffee that had scored the highest, that had extended the dream.

Mokhtar envisioned flying to New York, and then to London, and then to Sana'a, and after driving to Haymah, through the valley untouched by time, he would find Malik under his tree and tell him the news, that his coffee was among the best in the world.

CHAPTER XXV

A COUNTRY WITHOUT
A GOVERNMENT

THE CHILDREN HOLDING AK-47S—this was new. Mokhtar landed in Sana'a on October 27, 2014, and was confronted with the patchwork of overlapping military units, security forces and ragtag groups of Houthi or pseudo-Houthi rebels all over the airport and the roads to the capital.

How did a small group of hillbillies take over the country? Mokhtar had lived in California most of his life and that was the corollary that came to his mind: it was as if some almost-unknown militia from near the Oregon border swept down and took over Sacramento, San Francisco, Los Angeles, all without any significant resistance. One day Yemen had been ruled by President Hadi, and the next he was on the run, and this northern rebel group, the Houthis—who had scarcely had any real influence on Yemeni politics before—was suddenly in control.

In September the Houthis took over much of the capital. They had induced the surrender or complicity of most of the Yemeni army forces en route. Because the Yemeni army lacked top-down command, was instead controlled by its officers, few of whom were reliably loyal to Hadi, the progress of the Houthis was largely unimpeded. The Houthis bribed some commanders, and commanders loyal to Saleh eased their path. And then the capital was under their control.

When Mokhtar walked through the airport, he saw the Houthis everywhere—heavily armed but in traditional dress, with turbans, daggers—all of them strangely coexisting with the regular airport security. Mokhtar got in a taxi and within minutes they approached a checkpoint manned by Houthi forces. Or they were dressed as Houthis. But they were children, no more than thirteen. One looked to be ten.

They motioned the taxi driver to stop, and he did, and Mokhtar watched as a bizarre game of pantomime began. The children pretended to be men and soldiers, and the driver pretended that he was not noticing or caring that the soldiers were children. The children asked the driver for his papers and his destination, and after a cursory inspection, they allowed him to pass.

The remarkable thing about the Houthis was that they were polite. Mokhtar had been hearing this from friends in Yemen before he returned, and now he'd seen it at the airport and would see it again and again his first few days in the country. The Houthis spoke graciously, and in general were more professional and considerate than the usual authorities—more efficient and hospitable.

The taxi was stopped at more checkpoints on the way to the city, some manned by Yemeni police, some by Houthis, and all along Mokhtar hoped the car would not be searched, that he would not be searched. He was carrying ten thousand U.S. dollars, and was sure it would disappear if discovered. His investors had spotted him this much in cash, with the rest contingent on the conditions they'd agreed upon. For the time being, Mokhtar was happy not to have much paper money. Word would get around, and would make him too interesting to the Houthis or thieves. Paramount for him and his safety—and ability to do business in the tribal areas—was to remain uninteresting to all.

The business he hoped to do was not so complicated. He only needed to visit the coffee-growing regions whose coffee had scored highest, and then help to ensure that their upcoming harvests, in two months' time, were well monitored and that the cherries were picked at their carnelian-stone ripest. And then he had to buy about eighteen thousand kilograms of dried cherries—he would have to fill a shipping container with coffee. He had no funds to give them as a down payment, and they would be asked to forgo their usual buyers in hopes that a twenty-six-year-old from San Francisco—now he was twenty-six—could somehow secure, from unnamed investors, hundreds of thousands of dollars at some later date.

Then he would have to get these cherries to Sana'a, where the cherries would be hulled and sorted. But first he had to find or rent a processing plant. And then, if all else went according to plan—if he could buy the cherries, rent or buy a processing plant, and get the cherries there to process and sort them—then he had to figure out a way to ship eighteen tons of coffee out of Yemen, all during a civil war and while the Houthis controlled most of the ports.

Not so complicated.

Mokhtar arrived at Mohamed and Kenza's house a different man. They knew he had financial backing, at least in theory, and that he'd become a Q grader, which they knew meant a great deal in the coffee world, so they received him with a new kind of deference. He was no longer a student, and he wasn't a young man claiming to be starting a business. He was in Yemen to actually buy coffee, process it, store it and sell it in the international market. He had returned an Important Man.

But he was still sleeping on the floor. There was nowhere else.

With Nurideen, Mokhtar talked about the Houthis, the fact that Sana'a was under Hadi's control one day in September, and the next it was in the hands of the Houthis, while life in Yemen had gone on more or less uninterrupted. Banks and businesses were open the day of the invasion and open the next day, too. They talked about what effect, if any, the Houthis' advance would have on Mokhtar's work, and they concluded it would be minimal.

But now he would be traveling as an American exporter, carrying large amounts of cash to rural areas controlled by tribal forces. Considerations had to be made. He and Nurideen thought it through. He would need a driver, as usual, but this driver had to be armed. In some parts of the country he would need another guard, and this man should have an AK-47. Mokhtar planned to carry his SIG Sauer, which he usually traveled with in Yemen anyway, and now he'd wear a few grenades. (In Yemen, the grenades were more for show than anything. Men wore them on their chests, attached to vests, as a signal of their willingness to take any argument to its logical conclusion.) On any trip, then, there would need to be at least three guns in any vehicle, and if ever they were transporting coffee, they'd need multiple trucks, each with an armed escort.

*　*　*

The day after Mokhtar arrived in Yemen, Mokhtar and Nurideen went to Haymah. At the gas station called Abu Askr, they took the usual right and descended into the valley.

At the al-Amal Cooperative, Mokhtar got out of the truck and greeted all the farmers he knew. There were songs and handshakes and hugs, but Mokhtar was looking for Malik, the man he'd seen under the tree. He found him in a communal house, sitting with three other men. Mokhtar bent down and held Malik's head in his hands, and kissed him on the forehead.

"Your coffee is the best in the world," Mokhtar said.

Malik nodded. He said nothing.

"Thank you," Mokhtar said, and, given Malik's reticence, he felt the need to elaborate. He told Malik that he had brought his coffee all the way to San Francisco. That they'd cleaned it, sorted it, roasted it and cupped it, and that it had scored as high as any Yemeni coffee ever had.

Malik smiled and nodded.

"Thank you," Mokhtar said again, and told Malik that he would be buying all Malik's coffee from now on, at a price five times what he'd been paid before; that Malik's way of cultivation and picking would be the model for everyone in the collective; that together they would transform the Haymah Valley and, eventually, all the coffee of Yemen.

Malik nodded and smiled.

Mokhtar put his hand meaningfully on Malik's shoulder, and walked away. He almost laughed. Either Malik was a preternaturally stoic and unemotional man, or he had expected this news. Maybe, for him, just a confirmation of the obvious.

Mokhtar spent the day at the cooperative. He walked the farms. He talked about pruning and about the next harvest. He told the farmers about his Q grader status. They followed him through the trees, up and down the terraces, and, buoyed by the news of Malik's coffee, news that had quickly spread throughout the cooperative, they felt that Mokhtar might actually bring change.

No offense, they intimated in various ways throughout the day, but when he had first arrived, in his American clothes and unable to tell a coffee plant from an olive tree, they had been skeptical.

Mokhtar had never been to Ethiopia. He and everyone traveling the region had been through the Addis Ababa airport, but he'd never been beyond the city. The trip was the idea of the Small Microeconomic Promotion Service—SMEPS, an NGO with funding from the World Bank that sought to improve economic opportunities for small businesses throughout Yemen. The plan was to bring sixteen small coffee farmers from Yemen to visit successful farms in Ethiopia. Maybe they could take away inspiration and best practices. Mokhtar knew Abdo Alghazali, one of the directors of SMEPS, who invited him along. They flew to Addis Ababa, a brief flight over the Red Sea, on October 31, 2015.

For most of the Yemeni farmers, it was their first time out of the country, and certainly their first time in Ethiopia. A region near Harar had been chosen as a prime opportunity to see how specialty coffee was grown, harvested and processed on a large scale. From Addis Ababa, the drive to Harar was eight hours and cut through countless small towns. The views were magnificent—Ethiopia was green, everything along the way was green. Ethiopia, like Yemen, was a country plagued by misperceptions from the rest of the world. When the West thought of Ethiopia, they thought of poverty and famine, emaciated babies dying in the desert. But the Ethiopia Mokhtar saw was a bustling East African nation of cities and farms and lakes, with a large and educated middle class, a feisty press and, in Addis Ababa, a capital city that rivaled Nairobi and Johannesburg.

But they were not staying in Addis. They were passing through, on their way to Harar, the birthplace of coffee. It was in the hills of Harar that the mythical shepherd Khaldi had first noticed a sleepless spring in the steps of his goats, and had sampled the coffee cherries he'd seen them eating. There was still coffee grown in this region of Ethiopia, the Yirgacheffe, in vast hillside farms blessed by ample seasonal rain.

Harar, though, was something unique. An ancient city, home to some of the country's oldest mosques, a city almost untouched by modern architecture. In all of Ethiopia it was the most Yemeni city, a place where Arab traders had been coming for a thousand years and where they still exerted great cultural influence. Harar was also the adopted home of Arthur Rimbaud. The young French poet, who became a central influence for the Surrealists, exiled himself in a ramshackle home high above the city. A drug addict and sometime gunrunner, he was also, for a short period, a coffee trader. He died in France in 1891, at the age of thirty-seven, while planning his return to Africa.

The Ethiopian way of making specialty coffee was a revelation. Now the methods and standards Mokhtar had been preaching to the Yemeni farmers could be seen in practice. If they hadn't believed him before when he showed them pictures of vast drying beds full of bright red cherries, now they saw it for themselves. It could be done.

And done without vast sums of money or advanced technology. The Ethiopians picked cherries the same way the Yemenis did, by hand, but with greater care, and they used more precise methods all along the chain of production. The only facet of Ethiopian production that had no bearing on Yemeni growers was the use of wet processing. The Ethiopians were using vast amounts of water to wash their beans. Typically the coffee farms were located near rivers, and the Ethiopians were diverting river water to clean their coffee, and then allowing the runoff to re-enter the water system.

But the runoff was not potable, and because the water was now mixed with sugars from coffee plants, it altered the chemistry of any river or stream or water table it entered. In a world concerned with water usage and the dwindling access to fresh water—and the rising cost of fresh water—using so much of it in coffee processing wasn't, it seemed, politically or financially tenable in the long term.

Some Ethiopian coffee farmers were already experimenting with dry processing. In Yemen, Mokhtar knew they had no choice. Producers there knew no other way, and would rarely have access to the volumes of water necessary for wet processing. Dry processing had been the only way of Yemeni coffee since the beginning. It was both the strength of Yemeni beans and their weakness. Traditional dry processing had the potential to capture unusual flavors and could bring out the wildest and boldest parts of a bean. But dry processing, if done without great care, led to quality so inconsistent that the coffee was commodity-grade at best.

*　*　*

In Ethiopia, Mokhtar saw vast aboveground drying beds filled with ruby-red cherries. He saw small farms with their own varietal labels, farms sending their coffee directly to roasters in Europe and Japan. He saw the effects of direct trade, where the roasters told the farmers what they needed, and the farmers knew how to accommodate those needs. It was a beautiful symbiosis, without layers of brokers and loan sharks that invariably sapped the producers of profit.

Mokhtar flew back to Yemen, wanting to share what he'd seen with Yusuf and the al-Amal Cooperative. He tried calling them for days and got no answer. Finally Yusuf picked up.

"I'm sorry," he said. "We had a death in the village."

"Who was it?" Mokhtar asked.

"It was Malik," Yusuf said. "He died the night you left."

Mokhtar couldn't make sense of it.

"He was very old," Yusuf said.

Mokhtar went to Haymah to pay his respects. He found Malik's widow, named Warda, sitting in the upper story of her house. From the open windows, a cool breeze passed through the room. On the roof above, red cherries were drying. Mokhtar told Warda how sorry he was. Like her husband, she was very quiet, difficult to read. And like Malik, she was tiny, no more than five feet tall.

"I will take care of you," Mokhtar said. He told her how much her and her husband's coffee meant to him, and how he would always support her.

She didn't seem to have any idea what he was talking about. Mokhtar saw the scene through her eyes—her husband of fifty years had died a few days earlier, and now an American, whom she'd never met, was promising to take care of her?

Mokhtar met her son Ahmed, and they talked about the future. Mokhtar was conflicted, though. His own business depended to some extent on the ability of Malik and Warda's farm to continue to produce the quality of coffee that had scored so high just a month ago. It seemed doubtful that the farm could continue without Malik.

The General wants to see you. This was the message relayed to Mokhtar. He walked to the General's farm. From the start, the General had been the most suspicious of Mokhtar, given his Rupert clothing and city ways, but the sharpshooting contest had softened him to some extent.

Now they sat, just the two of them, and chewed qat. Mokhtar showed him photos from Ethiopia, the red cherries and drying beds, and the General examined them closely. The General, qat brightening his mood, talked about his time in the army and wanted to know how it was that Mokhtar had learned to shoot that well. Mokhtar told the truth, that he had learned in Bakersfield with Rakan and Rafik, and in Ibb, with Hamood. He mentioned that Rafik had been a cop in Oakland and had been the top marksman at the police academy. Somehow, over the next few weeks, this story evolved, traveling and expanding from the General throughout the village, until it was known that Mokhtar was the best marksman in California, having been trained by a Special Forces soldier.

The General pledged his commitment to Mokhtar's work, and promised to build the first drying bed in Haymah. A few weeks later, he had done it. The Ethiopians' version had been welded from aluminum, but otherwise the General's drying bed looked identical. It was enormous and sturdy and would hold ten thousand cherries, the bulk of his harvest. He'd built it from local wood after simply looking at the picture on Mokhtar's phone.

Hubayshi didn't call Mokhtar often. Usually it was Mokhtar calling him.

"I have twenty tons for you," Hubayshi said. "We picked them like you asked. Everything red."

Mokhtar was skeptical. Hubayshi was almost eighty and had been trading in low-quality commodity coffee for fifty years. Mokhtar had given him guidelines to meet specialty coffee standards, but he had no expectations that the old man would be able to get there—or that he would even try. Now he was saying he had twenty tons of specialty coffee.

When Mokhtar arrived the next day, it was true. Hubayshi's staff had picked the cherries ruby red and had kept the lots separated. They had bagged and labeled the coffees as Mokhtar had instructed. There were three primary sources: the Huwaar Valley in Ibb province, the village of Rawaat in the Udain region, and coffee from Wadi al-Jannat, the Valley of Paradise. In all, twenty tons. This was far more than the al-Amal Cooperative could possibly muster.

If Hubayshi's coffee cupped well, and if Mokhtar had the money to actually buy it, he'd have enough specialty coffee, eighteen thousand kilograms, to fill a container.

*　*　*

As Mokhtar traveled throughout the Ibb Valley, his entourage grew. There was always Nurideen, but now Yusuf from the al-Amal Cooperative often came along, as did a rotating array of other farmers he'd already convinced to join his movement. And there was no one more committed than the General. He loved field trips to other farms, and his presence was particularly crucial in convincing other villages, other co-ops, to adopt Mokhtar's methods.

One day, in a small village a hundred miles from Haymah, the entourage had taken a tour of the farms, and had eaten lunch, and afterward, about twenty local men were relaxing and chewing qat. Mokhtar, perhaps too emboldened by the qat, was expounding about not just the ways the farmers could and should improve their cultivation methods, but how they were currently being exploited, enslaved even, by the loan sharks who operated in the region.

"They take advantage of you," he roared. "You're selling far too low. Sell to me instead and you'll be free of these criminals. You'll be free, period. You won't be forever indebted to these sharks."

Mokhtar usually made it a point, whenever talking to an audience of farmers, to know the important players in the room, the head of the co-op and which elders held sway. But this time his intelligence was poor. The man sitting next to him, wearing a checkered kaffiyeh and carrying a ragged mix of paper and cash in his vest, was himself a loan shark—the very loan shark who had all the local farmers under his thumb.

He stood and turned to Mokhtar. "How can you come in here and tell these people this nonsense?" He narrowed his eyes at Mokhtar.

"You know, a man like you came here a few years ago. He was from Saudi Arabia, and came promising the same kinds of things. And he didn't come to such a good ending."

Mokhtar registered the message; this man was threatening to kill him. He reached slowly for his SIG Sauer, hidden in his sarong. He had no intention of firing it, but thought he might need it to get out of the village alive. He couldn't gauge the mood of the room. Were they aligned with the shark, or with Mokhtar?

Across the room, another man stood up. Mokhtar put on his glasses to see who it was. It was the General, his eyes enraged. He plucked one of the grenades from his jacket and raised it high over his head as he strode across the room, toward Mokhtar and toward the loan shark next. He put himself between Mokhtar and the man, the grenade almost touching the man's trembling face.

"If you interfere with Mokhtar," the General hissed, "you answer to me."

The loan shark smiled stiffly and sat down.

CHAPTER XXVI

MONEY IN YOUR HAND, NOT IN YOUR HEART

MOKHTAR WAS BUYING COFFEE but had nowhere to process it.

The only processing plant Mokhtar knew of was run by a man named Arafin Zafir. He'd met him months ago and knew his reputation. Zafir was Indonesian, and his Arabic was terrible, but he told his buyers in Asia and Europe that he was Yemeni. There were hundreds of thousands of immigrants in Yemen, and most of them had assimilated to some extent, but there was something about Zafir's way that bothered Mokhtar. And he processed his coffee in the same factory that made paper, a situation that prevented any of Zafir's coffee from being more than passable. There was no way to remove the vague smell of paper from the beans.

But for now Mokhtar had no choice. Abdo Alghazali had told Mokhtar not to work with Andrew Nicholson, the only other man he knew with a mill in Sana'a. His reasons for shunning this Andrew Nicholson—who had appeared in Mokhtar's original SWOT chart—were vague, but Abdo Alghazali was insistent. There would come a time when Mokhtar might have his own mill, but for the time being he had to choose the lesser of two evils, and that was Zafir. Mokhtar went to Sana'a with samples of all three Hubayshi lots.

At Zafir's plant, Mokhtar had to contend with Suha, who handled the day-to-day operations. Mokhtar gave her Hubayshi's samples, and put in an order to have the samples hulled and sorted. Suha was always haughty and usually brusque, and as Mokhtar talked to her, he saw the sorters at their posts, about twenty women sitting at wooden tables, their beans in two piles in front of them. The sorters were silent, prohibited from listening to music, and he felt for them then and every time he visited in the next week, coming back three different times to find his samples unfinished.

Suha provided excuses, but Mokhtar was running out of time. He needed the samples processed so he could roast and cup them, and send samples to Willem, who was in Ethiopia. But Suha was dragging her feet, and one day, amid the twenty silent sorters, Mokhtar lost his temper.

"If you can't run this mill," he roared, "you should sell it to me!"

Mokhtar had no idea why he said it. He had no money to buy a mill. But sometimes he put on the costume of the wealthy Yemeni American, knowing the people in a room like this wouldn't know if he was bluffing or not. The sorters looked up briefly before going back to work. But once Suha left the room, one of the sorters, an uncovered woman of about thirty, approached him.

"If you're buying this mill, take me with you," she said in English.

Her eyes were steady. Mokhtar was startled. The presence of an English-speaking sorter at a coffee mill was unexpected in itself, but

196

her boldness in speaking to him in front of the rest of the sorters was remarkable.

"I will," he said in English. "What's your name?"

"Amal," she said.

"Where can we talk?"

They arranged to meet the next day in a coffee shop.

When they met, she told him of the wretched conditions at Zafir's mill. The hours were long, the pay was dismal and it rarely arrived on time. They were not allowed to talk, to sing, to play music. One of the women had worked through the early stages of a pregnancy, and when she miscarried and fell ill, she was fired. Mokhtar thought of his grandmother in Richgrove, her many stories of the injustices visited upon farmworkers in the Central Valley. His grandmother's sense of outrage was his own.

"If you set up your own processing plant," Amal said, "I'll follow you. And I'll get the rest of the women to come with me."

Over the next two days at Zafir's, the women finished sorting his samples, and in a rush he packed them to be sent to Willem in Addis. He went to the DHL center in Sana'a and encountered the kind of add-ons he expected in Yemen. He had weighed his samples repeatedly and knew that he was sending Willem three samples, a total of 3 kilos. But the clerk at DHL said it was 4.2 kilos, and wanted a hundred dollars extra for it.

"Please," Mokhtar said. "Don't do this. I know it's three kilos."

The clerk weighed the package again, and again the digital readout said 4.2 kilos. Mokhtar checked to see if the clerk's hand was on the scale, but saw the man's hands at his side. Mokhtar had witnessed minor frauds like this dozens of times in Yemen, but this was impressive. Somehow the clerk had rigged the scale, he figured.

Mokhtar took the package off the scale and put it back. Again the weight was 4.2 kilos. Now he was curious. He opened the box, and then opened all three bags. The first looked the same. The second was unchanged. But when he opened the third, he saw something shiny and black. It looked like a SIG Sauer handgun. *His* SIG Sauer handgun. In his rush to get the samples ready, Mokhtar had thrown his handgun in one of the bags and almost sent a loaded weapon to Ethiopia.

Willem cupped the samples and found two of the three to be excellent. The Huwaar Valley sample scored 88.75, and the sample from Udain scored 89.5. The Valley of Paradise sample was overfermented, though, and was considered inferior.

But it didn't matter. Because Hubayshi had followed Mokhtar's guidelines, now Hubayshi was sitting on ten tons of high-quality beans from Huwaar Valley and seven tons from Udain. And Mokhtar knew he could buy it all. It would cost about two hundred thousand dollars to buy all eighteen thousand kilograms of dried cherries, and with Willem's help it would not be difficult to make a profit selling the container to specialty retailers in the U.S., Europe and Japan. Hubayshi already had trucks and drivers and knew how to move coffee through the country. All Mokhtar had to do was pay for the beans, process and sort them.

But when he called his investors, certain they'd share his enthusiasm for the high scores and the volume available to them—immediately, he added—they were unmoved. They were concerned about the security situation in the country, they said. Yemen seemed about to implode.

"So?" Hubayshi asked Mokhtar. He was calling daily.

"Just waiting for the funds," Mokhtar lied. "Any day now."

Every morning Mokhtar called and pleaded with his investors to actually invest in the start-up he was there to start up, and every day Hubayshi called, wanting to know if Mokhtar was going to pay for the coffee he had committed to buying. Hubayshi was gentle about it, but as the weeks went by, Mokhtar knew he would lose the coffee, one ton at a time. Hubayshi had farmers and collectives to pay, so he sold five tons of the Udaini and then five tons of the Huwaar Valley. Seeing the beans slip away drove Mokhtar to desperation.

In a symbol of goodwill, Hubayshi gave him five tons of the Udaini. He accepted the ten thousand dollars, and the rest of the cost—almost a hundred thousand dollars—on credit. That was not a problem. Mokhtar was sure he would eventually get his investors to buy the coffee they had sent him to buy, but for the time being there was the issue of where to put five tons of coffee.

Mokhtar had no warehouse and he had no mill. He didn't want to process his beans at Zafir's mill, given the working conditions. But the only other mill Mokhtar knew about in Sana'a was Rayyan, the operation run by Andrew Nicholson. Nicholson had been the first American that Mokhtar had located in Yemeni coffee. Despite Abdo's warnings, Mokhtar had no choice. Hubayshi needed his beans moved, and Mokhtar needed a place to process them.

CHAPTER XXVII

THE AMERICANS

WHEN MOKHTAR ARRIVED AT Andrew Nicholson's mill, his right-hand man, Ali al-Hajry, shot a rifle in the air. The mood was celebratory. Mokhtar greeted Ali and Andrew, and inside, the atmosphere was a perfect inversion of what Mokhtar had been led to believe. The workers were content and friendly. The sorters were singing. Almost immediately Mokhtar realized that Abdo Alghazali had wanted to keep Mokhtar and Andrew apart not because Andrew was an unscrupulous operator but because he knew that Mokhtar and Andrew would get along and that together the two of them would be unstoppable.

Andrew spoke Arabic with an accent specific to Sana'a. For a second Mokhtar and Andrew couldn't decide which language, Yemeni Arabic or American English, they should use. They decided on English, and Mokhtar heard a drawl from the American southeast. It was incongruous, even comical, coming from the mouth of a man in a sarong, with a Yemeni beard and a highly convincing Yemeni dagger tucked into his belt. He looked about as local as Mokhtar.

But Andrew had grown up in rural Louisiana. He played baseball and married his high school girlfriend, Jennifer. He studied engineering in college and afterward went into sales. Successful but restless, Andrew decided to go back to school, to get into nursing. A few years later, while serving as a nurse at a Houston hospital, he worked with doctors and other professionals from the Muslim and Arabic-speaking worlds, and became intrigued. This was after 9/11, and perhaps reacting against some of the bigotry he'd seen and heard as a child in Louisiana, he found himself drawn to his colleagues from Egypt and Jordan. If nothing else, he wanted them to know they were welcome.

Soon Andrew and Jennifer had decided to move to Yemen to study Arabic. They were in their twenties and more or less unencumbered. They didn't own a house, and though they'd just had their first child—she was nine months old when they moved to Sana'a—it was an adventure they felt they could only do at that liberated stage of their lives. After eighteen months in the capital, they had made friends and were fluent in Yemeni Arabic. They moved back to Houston, where Andrew became a consultant to companies operating in the Arab world.

One of Andrew's friends, Sean Marshall, owned a café in Houston, and he introduced Andrew to third-wave specialty coffee. One day they were talking about coffee, its origins and the state of the market, and Sean said, "What if you went back to Yemen and got some samples, and maybe tried to export Yemeni coffee?"

Andrew laughed it off, but in the morning the notion seemed more plausible. Andrew and Jennifer talked it over, and six months later, they moved back to Sana'a. They stayed with friends at first, and every week Andrew drove into the mountains to visit coffee farms and gather information. He returned to the capital with samples, but realized there was no one in Yemen who could properly process the cherries. He had no intention of starting his own processing mill, but without one, he had no business. So he worked with Sean and another partner to expand the scope of the operation. They'd work with farmers, bring beans to Sana'a, process them and export them. They called the company Rayyan.

The scale of the investment was steep, and Rayyan wasn't profitable in its first year, or second. Andrew couldn't find reliable employees, but he brought in one good man, Ali al-Hajry, who became deputy director of the mill. Everyone else Andrew hired stole from him. Andrew turned to Ali, and Ali turned to his mother. Ali's mother went back to her own village, twenty minutes from Sana'a, and put the word out: they were looking for reliable employees who wouldn't steal. In a few weeks, Ali's mother filled out the Rayyan staff. She knew everyone working with her son, who was working with the American. They were covered.

Rayyan began operations during the Arab Spring in 2011, but the chaos on the streets did little to diminish the company's early success. People wanted Yemeni coffee, and Andrew found himself exporting to Japan, China, Europe and North America. The upheavals in Yemen inconvenienced their work occasionally, but Rayyan managed to operate continuously through the rise and fall of President Hadi and the arrival of the Houthis. It was all to be expected when running an export business from Sana'a.

Mokhtar and Andrew agreed that they would work together, as partners and not competitors. Mokhtar would seek out high-end coffee in the hinterlands, while Andrew would stay closer to Sana'a and concentrate on exporting more affordable coffee. Rayyan would process Mokhtar's beans, but Andrew couldn't do the sorting. He didn't have the space or the staff.

The ground-floor retail space of Mohamed and Kenza's building, that had until recently been a corner store where candy and soda were sold, was empty. Mokhtar thought it would make a convenient location, but Nurideen was dubious. The location was convenient, sure, but there was the matter of the demon child who had been sighted there the last time the space was open.

It had happened just a few months ago. In the middle of the day, a thirteen-year-old child had been found standing in front of the store with a knife in his hand, rolling his eyes and speaking in tongues. No one could reason with him, and finally the consensus was that he was possessed by an evil spirit. The boy was taken to an exorcist in Sana'a, who deduced that there was indeed a demon inside him, and that this demon was in love with the child. Given the taint the devil had conferred upon it, the store was closed. Nurideen related all this to Mokhtar.

"We're talking about the store on the first floor, right?" Mokhtar clarified. He'd been inside a hundred times; he used to buy his phone cards there. Mokhtar didn't believe that a demon had possessed the child and certainly didn't believe that the storefront itself was tainted as a result. But the demon-taint had left the storefront empty and the rent affordable. So he leased the space.

"But don't tell the sorters," Nurideen warned.

Yemenis are superstitious, and if even one of the sorters was spooked by the story of the demon child inhabiting the storefront, it would be incentive enough for the rest of the sorters to stay at Zafir's demon-free mill. So Mokhtar told them nothing.

He rented the space, and also rented the two adjoining storefronts, and knocked out the walls between. He created a lounge area with couches, a coffee table and carpet. He doubled the pay for Amal and the other sorters, and one day in February of 2015, sixteen sorters left Zafir's operation and arrived at Mokhtar's.

Mokhtar made an event out of it. He had never employed anyone, but had picked up a few notions from the progressive California companies where his friends worked. He had in mind a kind of staff-orientation day. He provided coffee and juice and cakes and gathered all sixteen women together, asking them to sit in a circle. They all wore niqabs. He could see only their eyes.

"I want to know about each and every one of you," he said, and saw in the women's eyes that this was highly unusual. "Let's go around the circle, and each of you can tell me your name, where you're from, and just as an icebreaker, you can tell me a food that represents you and why."

The women had no clue what he meant. Why would they be a food? Why would an employer want to know such a thing? It took him twenty minutes to explain the concept. Finally, he got one of the women to suggest that if she were a food, she'd be a green apple. She was named Um Riyadh, meaning "mother of Riyadh." She was the oldest of the group, and Mokhtar could tell she was bolder, more outspoken, than the others.

"Why a green apple?" Mokhtar asked.

"Green apples can be both sweet and sour," she said. "And I'm the same way. Sometimes I'm sweet. Sometimes bitter. It depends on my mood."

The other women laughed tentatively.

"Good, good!" Mokhtar said.

But when it was the next woman's turn, she said the green apple represented her, too. The next woman said the same, that she, too, was a green apple. The women still didn't understand the concept, and chose to copy one another rather than venture into the unknown.

But he did get their names and hometowns, and he startled them all by knowing something about each of their origins. They expected him to be ignorant of the lesser-known regions of Yemen, but after visiting all thirty-two coffee-producing areas of the country, he knew the land as well as anyone.

Ahlam said she came from Utmah.

"I've been there," Mokhtar said. "Amazing guavas there."

Um Riyadh said she came from Bani Ismail.

"I've been there," Mokhtar said. "You have those little monkeys that roam around in packs."

Baghdad said she was from Haymah.

Mokhtar asked her if she came from inner or outer Haymah. She seemed dubious that he would know anything about Haymah. She said she was from outer Haymah.

"Al-Mahjar?" he asked.

"Lower," she said.

"How about Bait Alel? Bait al-Zabadani?" he asked.

"Almost there," she said.

"Al-Asaan?" he guessed.

"Yes!" she said.

The room broke open with cheers.

Mokhtar unfolded his laptop and showed them pictures of his journeys across Yemen. They huddled close, disbelieving. They had no idea how diverse Yemen was, how beautiful.

After a few hours he had a sense of them, though he saw no more than their eyes. At Zafir's plant, the sorters' area was open to the larger factory, so the women there had to remain in niqabs all day, an uncomfortable and impractical state of affairs, given the labor of sorting and the lack of air-conditioning.

Mokhtar was determined to fix this. He couldn't operate a company in a busy Sana'a neighborhood with sixteen women uncovered and visible to passersby—the customs of Yemen frustrated him as much as anyone, but he couldn't risk the entire business to make a point about traditional clothing for Yemeni women. Compromise was necessary for now. He rearranged the space such that there was a large room with high walls and a door that closed and locked from the inside. The sorters could control who entered and when, and when they were alone, they could take off their niqabs and dress and act as they pleased.

Mokhtar thought of the policies he'd wanted to change at the Infinity. He provided free breakfast and lunch. Free wifi, transportation to and from work. He gave them May Day off and a sound system they could connect to their smartphones.

"Do whatever you want to do while you work," he told them. "You are my golden team." He felt benevolent and thought that in their little space, they might approximate a Californian kind of operation—liberalized and egalitarian.

But for the first few days, as he trained them, they were ill at ease, and though he'd provided the sound system, every time he walked by their workspace, he heard silence. He added a living room, with couches and a prayer corner. Still nothing.

A week in, though, he heard something. He was on his way to Rayyan when he heard a deep-bass thumping coming from the sorting room. The women had closed their door and locked it from the inside, and Mokhtar stood by the door, feeling sure he recognized the song. It was Usher's "Yeah!"

From then on, every day there was music coming from the sorting room. Sometimes it was traditional Yemeni music. Sometimes it was Katy Perry—in particular, they loved "Roar." Often they sang along.

"You're in front my face," he told them.

He'd said it the first day, and he said it whenever they had a meeting, whenever he needed to remind them of their importance to him. In following him, they'd risked a great deal and he would not forget it.

"You're in front my face," he said every day.

It was an old Yemeni expression, hard to translate into English. It was something you said to a loved one, to a friend, while pointing to your own face. It meant that the person before you was never out of your vision. That you kept them foremost in your mind.

BOOK IV

CHAPTER XXVIII

BEDLAM

IT WAS DECEMBER 31, 2014, three days before the Prophet Muhammad's birthday. There were celebrations all over the country. Mokhtar woke up in his grandfather's house in Ibb with a plan to work out in the gym around the corner. But when he went downstairs for breakfast, he found his aunt staring at the television. A suicide bomber had struck Ibb.

Forty-nine people had been killed and seventy wounded. It was the first attack of its kind in Ibb, a city accustomed to being distant from this kind of violence. *It's begun,* Mokhtar thought. He had to be in Sana'a that day, so he left Ibb and drove north.

When he arrived at Mohamed and Kenza's house, there were urgent discussions about the implications of a bombing of innocents in Ibb. Even al-Qaeda wouldn't have perpetrated such an attack in Yemen. It typically directed its fury at Western targets, military targets—not Yemeni civilians.

* * *

A week later, Mokhtar was back at Mohamed and Kenza's. One morning after breakfast, he decided to work out. The closest facility was called Health and Sports Club Arnold, named for Schwarzenegger.

He took a taxi, and a few blocks from the gym, he noticed a crowd of men walking into the neighborhood. They were mostly Houthis. The gym, he realized, was close to the police station, and today was a recruiting day for the police academy. Part of the Houthis' plan was to stack their members in the police ranks.

Mokhtar paid the driver and decided to walk the few blocks to Arnold's. He wanted to get a sense of what was happening, taking in the strange sight of so many northerners in the center of Sana'a.

Then the earth convulsed. He fell to his knees. He thought it was an earthquake. He felt the concrete, expecting vibrations, aftershocks. He heard screams. The wail of car alarms. He ran toward the police station and saw the burning flesh. A man's torso lay on the ground. A woman was screaming. Blood stained the street. He scanned the charred remains of dozens of dead, thinking he saw people he knew.

With a start he remembered that Hathem, Mohamed and Kenza's second-oldest son, was a cadet at the academy. Mokhtar knew he was home that day, that there was no way he would be among the dead, yet he saw Hathem's face in the charred husks of men. Soon newspeople arrived. They began filming and taking pictures. Mokhtar put his phone away. He couldn't look at the carnage anymore. And then he had a thought: *Run.*

He ran. He knew terrorists often set off a second bomb once rescuers had arrived to treat the wounded from the first. So Mokhtar ran, and everyone around him, thinking he'd seen something, ran, too.

At Mohamed and Kenza's house, he said nothing. Nurideen knew something was wrong, but Mokhtar didn't want to burden the family with what he'd seen. They would see the footage on television. Thirty-eight had been killed, sixty had been wounded. Mokhtar didn't want his family to worry. But privately he wondered what was happening in Yemen and worried the country was going the way of Iraq—a lawless land of sectarian strife, suicide bombs, kidnappings, and the impossibility of untroubled life.

Mokhtar went about his work as best he could. Every day he opened the processing mill for his sorters and provided breakfast, and they talked about the previous day's work and what needed to get done that day. He continued to train them, watching each of them, giving them notes.

After lunch he'd go to Andrew's mill to chew qat and talk business and politics. They didn't know if anything happening in Yemen would materially affect the country or their work. It seemed, at first, simply the usual transfers of power between equally incompetent players.

On January 7, two brothers, Chérif and Said Kouachi, entered the Paris offices of the satirical magazine *Charlie Hebdo* and shot eleven people dead. On their way out, they shot and killed a policeman. A third man, Amedy Coulibaly, shot and killed a policeman in Montrouge, south of Paris, and four at a kosher market. The attacks, precipitated by *Charlie Hebdo*'s publishing of cartoons depicting the Prophet Muhammad, brought forth a global outpouring of support for the magazine and its slain writers, editors and cartoonists. On Sunday, January 11, more than four million people throughout France marched in support of the victims and the right to free expression.

On January 14, the Yemeni branch of al-Qaeda—al-Qaeda in the Arabian Peninsula—claimed responsibility for the attack.

On January 18, the Houthis in Yemen rejected a new constitution drawn up by committee across the political spectrum. The next day, they seized the state television station. They took over all the government buildings in Sana'a, and President Hadi resigned in protest. Weeks later he rescinded his resignation, but it didn't matter. The Houthis had control of the country.

Mokhtar's family and friends in the United States worried for him, but Mokhtar saw very little change in his daily life. He'd gone to bed one night with Hadi in power, and woke up without a president. And yet the airport was still open and hosted regular commercial flights. The banks functioned as usual. And the grocery stores, the health clubs, the mosques. Taxi drivers drove their taxis. Sana'a was still Sana'a, though it was now run by the Houthis. The lives of everyday working Yemenis continued unchanged. Mokhtar spent afternoons at Andrew's mill, chewing qat with Ali and together they laughed at the Yemeni Americans who were fleeing the country.

"We have no government now," Mokhtar said.

"Wait. Yemen had a government?" Andrew asked.

On February 10, 2015, the U.S. State Department announced that it was suspending embassy operations. The staff, it said, had been relocated out of Sana'a. The next day the U.S. embassy was closed for good. The British embassy closed, too, and the U.S. and British governments urged their citizens to leave the country immediately. But there were no plans for a U.S. evacuation of American citizens.

Commercial flights were still available, the State Department noted, and "U.S. government–facilitated evacuations occur only when no safe commercial alternatives exist."

The French foreign ministry closed a few days later. "Given the recent political developments, and for security reasons," their statement read, "the Embassy invites you to temporarily leave Yemen, as soon as possible, via commercial flights at your convenience. The Embassy will temporarily be closed as of Friday, February 13, 2015, until further notice."

It was not unusual for the Western embassies to close for a day or a week. The U.S embassy had closed in 2001, 2008 and 2009. It was part of the rhythm of life in Yemen, Mokhtar assumed. The heat turned up, the embassies closed, and then, a few weeks later, with things calm again, they would reopen.

Andrew was staying, too. They promised to keep each other informed of their plans. Until things became untenable—or, more specifically, until they couldn't continue their coffee work—they would remain.

A few hours every day, Mokhtar tried to convince his investors to pay for the coffee he'd promised to buy from Hubayshi. It had been two months, and the investors were intractable. And as conditions worsened in Yemen, their hold on their money became ever tighter.

Mokhtar called Ghassan for advice. He called Willem. And Hubayshi called him daily. *Where is my money?*

By March, most of the Yemeni Americans Mokhtar knew had left. He began to worry for Andrew. Mokhtar was part of a vast tribe and could count on their protection. But Andrew had no such heritage. In this new period, where standards and order were no longer reliable, any foreigner could become a target for kidnapping. Jennifer, Andrew's wife, would be safe if she stayed inside, but Andrew was known and would be sought out.

On March 20, suicide bombers detonated themselves inside two different mosques in Sana'a. Because it was during Friday prayers, the mosques, used by Houthi Shiites, were full. One hundred and thirty-seven men, women and children were killed, and over three hundred were wounded. This was the worst-ever terrorist attack on Yemeni soil, and was claimed by ISIS.

On March 21, ISIS posted the names and addresses of all one hundred American military personnel in Yemen and encouraged its acolytes to kill them. These last U.S. personnel were evacuated on March 25, and the same day they left al-Anad, just north of Aden, the Houthis quickly seized the strategic military base. The Houthis also took control of the Aden International Airport and Aden's central bank.

Saudi Arabia, which had clashed with the Houthis in 2009 and 2010, now massed artillery and tanks near the Yemen-Saudi border. By late March, the Houthis controlled nine of the twenty-one provinces of Yemen, including Taiz, the country's third-largest city.

"It looks bad," Mokhtar said.

He and Andrew were at the mill, chewing qat in the afternoon.

"It's Yemen," Andrew said.

They were not yet committed to leave Yemen for good. But they were planning to leave temporarily to attend the Specialty Coffee Association of America (SCAA) conference in Seattle. Rayyan had rented a booth, and with hundreds of importers and buyers there from around the world, Mokhtar and Andrew considered the conference a crucial step in their work. Mokhtar would share Andrew's booth and present his Haymah and Udaini coffees. The conference would be his most important event to date, his first real introduction of the progress he'd made with Yemeni varietals.

Getting out of Yemen even during peacetime, even for American citizens, had become a fraught proposition. The stories started circulating in 2011. Yemeni Americans would go into the U.S. embassy for some routine request and leave without a passport. There had been bizarre interrogations, accusations of Yemeni Americans changing their names, living in the U.S. under false identities. Mokhtar had heard the stories. They were strange and unreal, none more so than that of Mosed Shaye Omar.

He was an acquaintance of Mokhtar's from San Francisco. A gentle man of about sixty, he'd lived in the United States for over forty years. In 1978 he'd become a naturalized U.S. citizen. He had a social security number, a California driver's license, paid his taxes faithfully.

Like thousands of other immigrants, he'd left family members in his country of origin. In his case, one daughter had stayed behind with his own parents in Yemen. When she was twelve, Mosed was ready to bring her to live with him in San Francisco. In 2012 he went back to Yemen to prepare the paperwork to get her a passport. In August of that year, at the U.S. embassy in Sana'a, he submitted his application for a U.S. passport for his daughter.

In December of 2012, he was summoned to the U.S. embassy. They called him on the phone to say they had "good news" about his daughter's passport application.

On January 23, 2013, he went to the embassy, thinking he'd be

217

picking up his daughter's passport. When he got there, a consular official asked for his passport. Mosed handed it to the official, who asked him to sit in the waiting room.

About an hour later, Mosed was escorted out of the waiting room. He followed an official through the main building and into an adjoining building. They walked through a number of secure doors, guarded by uniformed U.S. military personnel. Already he knew that this was likely a departure from anything standard. He knew he was not being led to his daughter's passport.

He was taken into a small room where there were three men. One was an official with the Diplomatic Security Service, and another man served as an interpreter. The third man seemed to be an American, but he didn't speak throughout the process, which Mosed came to understand was an interrogation.

Through the interpreter, the man from the Diplomatic Security Service asked Mosed questions about his origins, his family, his name. Mosed told him his name was Mosed Shaye Mosed. It was the name on his passport, after all—a passport granted and renewed by the State Department as recently as 2007.

After an hour, Mosed was escorted out of the interrogation room and back to the waiting room he'd started in. He was told to remain there.

An hour later, he was brought back through the secure doors and hallways, past the armed guards again and back to the interrogation room. The man from the Diplomatic Security Service again asked Mosed about his name. Mosed again insisted his name was Mosed Shaye Omar, the only name he'd ever had. The interpreter seemed frustrated with Mosed and began inserting his own advice into the dialogue. Beyond translating what the Diplomatic Security Service agent said, he began telling Mosed to cooperate, to tell the agent what he wanted to hear.

This second session lasted another hour, at which point Mosed was again brought back to the general waiting room. He was told to wait there. Hours passed. Mosed hadn't had access to food or water since six-thirty that morning, difficult in normal circumstances but in Mosed's case especially trying. He suffered from diabetes and high blood pressure, and while waiting, he felt faint, and his vision blurred. He couldn't call family or friends, given cell phones were not permitted in the building, and the building had no pay phones.

At 4:00 p.m., Mosed was so desperate to be released that he approached a guard and told him that he needed to leave, that he would do anything to be allowed to go home and eat. The guard conveyed this to the embassy officials, and soon a consular official arrived and escorted him back to the interrogation room.

There, he was given a piece of paper to sign. He was not a fluent reader of English, so he couldn't parse the paper's meaning. The interpreter was in the room but didn't offer to translate it. Mosed was told to sign it if he wanted to leave the embassy. He signed the paper with his name, Mosed S. Omar, and the interpreter took his thumbprint.

After he signed the statement, he was taken back to the waiting room and told to wait for the consul, who would return his passport. But his passport was not returned. Mosed was called to the window and told that his passport could not be returned because his name was not Mosed Shaye Omar. At that, the official closed the window, left the room, and Mosed was escorted to the door by an armed guard.

Mosed went home, and because he hadn't eaten or had water in twelve hours, he experienced a severe diabetic attack, and was rushed to the hospital. As he was being treated, he struggled to understand what had happened at the embassy. He'd been given no explanation of how they'd come to think he was not actually Mosed Shaye Omar. They provided no evidence. They offered no explanation and gave him no opportunity for redress. He wasn't given a date for a hearing, or any idea of what he was supposed to do without a passport, while most of his family was in the United States, and he had lived there for forty years.

The next day, he began calling the embassy, but the phone was never answered. He learned that e-mail was the embassy's preferred method of communication, so he began writing e-mails. For the next eleven months, he wrote to the embassy but received no response. Finally, in December of 2013, almost a year after his passport was confiscated on January 23, he received an e-mail telling him to come to the embassy. He did so on December 15, and when he arrived he was given a written notice explaining that his passport had been revoked because "an investigation revealed that you are not Mosed Shaye Omar, born on February 1, 1951. In fact, you are Yasin Mohammed Ali Alghazali, born on February 1, 1951. On January 23, 2013, you signed a sworn statement admitting that your true identity is Yasin Mohammed Ali Alghazali. Because you made a false statement of material fact in your passport application, your passport is revoked pursuant to Section 51.62(a)(2) of Title 22 of the U.S. Code of Federal Regulations."

Mokhtar heard similar stories for years. Going to the U.S. embassy for any kind of help, then, was not an option.

On March 25, just after the final American troops abandoned Yemen, and a day after Houthi forces advanced on Aden and forced President Hadi to flee by sea, Mokhtar directed a taxi to his local Yemeni travel agent, whose office was still open in Sana'a, to buy plane tickets to Seattle, for a coffee conference.

But when the driver got close to the travel office, Mokhtar saw a throng of Houthi mourners. It was the funeral for the victims of the March 20 attack. Soon the crowd had spilled into the street and the taxi was surrounded. Mokhtar knew he shouldn't be there. The funeral was a target—terrorists had made a habit of bombing funeral gatherings to double their body count. Mokhtar got out of the taxi and pushed his way through the crowds. He'd get his ticket later.

The next day, President Hadi made a direct appeal to Saudi Arabia, asking for its help in turning the tide on the Houthi movement. Citing the Houthis' ties to Iran, he pleaded for direct Saudi military involvement. Mokhtar heard about Hadi's appeal but didn't think much of it. No one did. Mokhtar couldn't remember if the Saudis even had an army.

CHAPTER XXIX

MOUNTAINS ON FIRE

AT 3:00 A.M. ON March 26, Mokhtar was rattled awake. The building was shaking. He was at the Rayyan mill; he'd been working late and decided to sleep in Andrew's office. The rattle brought him to the roof, where he saw Faj Attan Mountain on fire. Houthi antiaircraft fire striped the sky. Fires plumed around the city. It was the end of the world.

Mokhtar went online and confirmed it was the Saudis. F-15s were bombing Houthi positions all around Sana'a. Every few minutes there was another strike. The ceiling shook and dust rained down.

Mokhtar called his mother. "I'm okay," he said. She begged him to leave the capital and go to Ibb, to Hamood's house. Mokhtar thought it through. Ibb was no doubt safer—the Saudis were unlikely to bomb Ibb. But going anywhere in the middle of a bombing campaign seemed unwise. Mokhtar was in a high-density residential neighborhood of Sana'a, and from all the news he was getting it seemed that the Saudis were after the Houthis' military positions and munitions dumps only. They wouldn't bomb a civilian neighborhood.

He told his mother not to worry and hung up. He tried to sleep. He counted the air strikes. Fifty, sixty. He lost track at eighty.

At 5:00 a.m., he heard the call to prayer. Then another. Competing calls echoed through the city. He went to the street, determined to wait out the last hour of darkness at the mosque. On his way, between the black silhouettes of the buildings, he saw the bright white stripes of antiaircraft fire.

Inside the mosque, a few dozen men were gathered as the bombing continued. The rug was gray with the ceiling's plaster. The imam performed a long supplication, and the congregants prayed as if living their last minutes. There couldn't be so many military targets in Sana'a, Mokhtar thought. They must be hitting civilians and this must really be war. When the imam asked God to forgive the sins of those present, the men around him wept, and Mokhtar knew he might die there, that at any moment a bomb would rip through the roof.

Had it been a good life? Mokhtar thought. He wasn't sure. It was incomplete. He should have started all this coffee business sooner, he thought. Had he begun a year earlier, he would have at least done something, finished something, before the bombs rained down. Now he would die in a mosque. Maybe his family might find some comfort in that. Another bomb struck, now closer.

The men around him stopped crying. They had submitted to their fate. Mokhtar did, too. Nothing was within his control, so he lost all fear and worry. He felt a weight leave his shoulders. He would die, he would not die. It had nothing to do with him. He could run from the mosque and die. He could stay in the mosque and die. He could go to Kenza and Mohamed's home and die among his family. He could go to Rayyan and die with his coffee.

Or maybe he wouldn't die. He and the congregants stayed an hour, until finally the quiet between bombs spread and became whole. At daybreak it was over. When Mokhtar and the rest of the men left the mosque, the sun had begun to rise and the city was bathed in an eerie pink light, the air bright with dust.

Mokhtar, feeling a new and encompassing peace, walked from the mosque to the mill, sure that nothing would ever frighten him again. It was as if he had died already.

Later that morning, he went back to the travel agent. He told her he wanted two tickets out of Sana'a. He and Andrew had to get to the SCAA conference. "What are you talking about?" the travel agent said. "There's no *airport*." The Saudis had destroyed runways and threatened to shoot down any plane leaving Sana'a.

Mokhtar went to the mill. He and Andrew chewed qat. "It was closed during the Arab Spring, too," Andrew said. "It'll reopen."

Mokhtar checked the U.S. State Department website, expecting to find information about an organized evacuation for American citizens. There was nothing of the kind. Every day, the State Department offered vague indications that Yemeni Americans should find passage out of the country by any means available.

There was recent precedent for the U.S. State Department helping its own citizens evacuate from a foreign country at war. In 2006, the Pentagon and State Department helped fifteen thousand

Americans leave Lebanon during the war between Israel and Hezbollah.

But this was different. Given the presence of AQAP and ISIS, the U.S. decided it could not risk a large evacuation. They had no embassy or staff on the ground, thus had no effective way of screening all the prospective passengers on a plane or ship. They deemed the prospect of unintentionally bringing a terrorist into the United States too great a risk. They decided to leave American citizens stuck in Yemen to their own devices.

An official notice from the U.S. State Department said that "there are no plans for a US government-coordinated evacuation of US citizens at this time. We encourage all US citizens to shelter in a secure location until they are able to depart safely. US citizens wishing to depart should do so via commercial transportation options when they become available."

This led to the creation of a website, StuckInYemen.com, which documented the plight of those remaining in Yemen. The site was supported by the American Muslim advocacy groups including the Council on American-Islamic Relations and the Asian Law Caucus. The site grew to include a registry of seven hundred Americans hoping their government would provide a way out of Yemen.

Under pressure from Arab American civil rights groups, another State Department spokesman, Jeff Rathke, explained that those Americans remaining in Yemen had made their own bed and now they must lie in it. Because they had ignored long-standing warnings from the American government, he implied, this was on them. "For more than fifteen years the State Department has been advising U.S. citizens to defer travel to Yemen, and we have been advising those U.S. citizens who are in Yemen to depart," he said.

At another State Department press conference, another spokesperson, Marie Harf, referred vaguely to escape "opportunities" for Americans.

One reporter asked her for clarity. "What are those opportunities?" he asked. "Swim?"

Mokhtar needed to get out immediately. He had to make it to the SCAA conference in Seattle, and he had to escape the escalating violence (in that order). Every day, he checked with the travel agent to see if any flights were leaving. But the airport was still in shambles. There was no hope of it opening anytime soon.

The bombing continued, most of it concentrated at night. The Saudis had named the campaign Operation Decisive Storm, and claimed to have the participation of nine other nations, most of them with predominantly Sunni populations. Jordan, Morocco, Sudan, Kuwait and Bahrain had supplied fifteen jets each. The United Arab Emirates had provided thirty. Senegal, Qatar and Egypt were part of the coalition, too. But most of the operation was Saudi led, with 100 Saudi jets participating and 150,000 Saudi troops mobilized.

The scope of the bombing expanded. First it was the air force base outside Sana'a and munitions dumps. Then major roads connecting the capital to Taiz and Aden. By Saturday, March 28, at least thirty-four civilians had been killed in the strikes.

*   *   *

By its fifth day, the bombing had become oddly normalized, at least in the center of Sana'a. Mokhtar went to Andrew's house. Andrew might have options, he thought, might have answers. He definitely had qat.

Mokhtar got in a taxi and directed the driver to El-Bonia. When they were close to Andrew's house, smoke billowed from the taxi's engine.

"Overheated," the driver said, and they stopped.

Mokhtar got out and saw a shop selling flower necklaces—a Yemeni version of a Hawaiian lei. He laughed to himself, thinking he'd buy one for Andrew and one for Ali. Some kind of bombing-survival gift. He bought two and got back into the taxi as the driver closed the hood.

Andrew and Ali laughed when Mokhtar presented the leis. Andrew put his on, and the three men sat in Andrew's apartment and stuffed their cheeks with qat. Mokhtar opened his laptop, looking for news. There was nothing promising. He went on the U.S. State Department site to see if there were any options. None. The afternoon wore on.

At dinnertime Ali said he'd drive Mokhtar home. The city was quiet. Knowing the bombing could start anytime after dark, the residents of Sana'a had gotten into the habit of choosing their destinations before night fell. No one wanted to be in the streets after that.

The sky grew dark while Mokhtar and Ali were still making their way across the city. And now Ali suggested going to the mill. He needed to be there, and it was on the way to Mokhtar's. Would he mind?

Mokhtar had no choice. All the taxis were gone for the night. They went to the mill. On the way, the bombing began. This was the first night Mokhtar was moving while the bombs fell, and the sensation was new. The ground rumbled beneath the taxi. There was the faint shushing of faraway targets reduced to dust.

When they got to the mill, they watched the war through the window. Antiaircraft fire lit up the sky. Mokhtar turned on his phone's camera and filmed the tracers as they flew upward beyond Faj Attan Mountain. The Saudis had hit a munitions dump. An orange cloud of fire bloomed to three hundred feet. There were explosions within the explosion. It was no more than a quarter mile away.

It's time to leave Sana'a, Mokhtar thought.

But this was no time to leave the mill—at least not that night. With this new attack so close to the mill, Mokhtar had no idea what might happen next. The Saudis had already hit homes, markets, hospitals, but there had been some ostensible plan to their bombing. Now it seemed possible they might begin bombing industrial buildings. There could be chaos, looting. Mokhtar thought of the five tons of coffee he had stored in the mill. If it was stolen, that would extinguish all the work of the last eighteen months.

"You know what?" he told Ali. "I'll stay here tonight."

Ali refused to leave him. "I'll take you home," he said. There was no logic in staying so close to this last explosion, he insisted.

Mokhtar told him to go home, that he'd stay and watch the mill.

Ali left, and Mokhtar set himself up in the office. He gathered the couch cushions and fashioned himself a bed. The bombs shook the city every ten minutes, but he grew used to it. Just before midnight he found himself drifting off to sleep.

His phone pinged. *Don't check it,* he told himself. *Just try to sleep.* He checked it.

It was a message from Summer Nasser. Mokhtar knew her through social media; she was a Yemeni American based in New York. While visiting family in Aden, she'd gotten stuck in Yemen, too. She'd heard there was a Greek ship leaving Aden in the morning, at 9:30 a.m. She was going to get on it.

"I'll save you a spot," she said.

CHAPTER XXX

THE SUMMER SHIP

MOKHTAR WAS SUDDENLY VERY awake. Aden was an eight- or nine-hour drive, due south. He'd have to find a vehicle that could make the trip. And a driver. Maybe a bodyguard. They would be traveling through an active war zone. There could be dozens of checkpoints. He'd have to pack a sampling of his beans, and enough cash to get on the ship and onto a flight to Seattle. They'd have to drive double time through the heart of Yemen, in the middle of an all-night Saudi bombing campaign. It was a ludicrous proposition.

But then Mokhtar had an equally absurd vision of himself in Seattle, speaking to coffee buyers there, telling them the story of Yemeni coffee, collecting orders, preselling tons of coffee, making this business real. He wanted that. And he didn't want to let Saudi bombs determine what he could and couldn't do. He prayed the Istikhara, a prayer to God to provide answers.

Is this the right path? he asked Allah.

He felt an answer: it was right.

That was enough. He wanted to go, and Summer's message appearing just after he'd seen a nearby mountain explode: it seemed like a confluence of meaningful indicators. The last time he'd had this same feeling—a feeling of destiny obliterating all doubt—was when he'd seen that Hills Bros. statue across from the Infinity and decided to devote his life to coffee.

He called Summer. "I'm coming."

He called Andrew and told him about Summer and the Greek ship. Andrew was half-asleep.

"Don't go," he told Mokhtar. "Aden is an *actual war zone*. There is an *actual ground war* happening there."

Mokhtar was undeterred. Andrew called Ali.

"Can you talk some sense into Mokhtar?"

Ali called Mokhtar, but Mokhtar couldn't be dissuaded. Finally Andrew and Ali gave up trying to stop him, but they wouldn't let him go alone.

Mokhtar called his family's driver, Samir. Mokhtar asked him to go, told him he'd be well paid. Samir was terrified.

"No," Samir said. "And you shouldn't go either."

Mokhtar hung up. He had no options.

Meanwhile, Ali called two friends, Sadeq and Ahmed. They lived in the neighborhood of Andrew's mill and had helped the night before, when Andrew moved his beans from the mill to his house. They agreed to make the drive to Aden for a modest fee. Sadeq said he could borrow the truck he usually drove during the day. He didn't own it, but the company he worked for wouldn't know. Mokhtar negotiated a price for the vehicle and for Ahmed to drive through the night, Sadeq along for numbers.

* * *

Mokhtar started packing. What did he need? He rushed back to Kenza and Mohamed's apartment and put two clean shirts and a pair of pants in a backpack with his phone and laptop. He added a change of socks and underwear. He strapped four thousand dollars in U.S. currency to his waist and tucked his Colt .45 into his belt.

Now the beans. He borrowed a hard-shell Samsonite suitcase and went downstairs to his sorting room. He grabbed a bag of beans from Haymah. The widow Warda's. The General's. Hubayshi's. What else? He saw the faces of the farmers. How could he leave any of their work behind? He settled on an assortment from the north and south, from six different farms. Whatever he brought would have to represent Yemeni coffee in Seattle. A sampling of the best beans he had—the best beans grown in Yemen in eighty years, a haphazard but still significant representation of the coffee from the country where coffee was first cultivated and the manifestation of five hundred years of tradition.

Mokhtar closed the suitcase and tried to lift it. It was too heavy and wouldn't zip. He had to lighten it. What segment of the history of Yemen could be removed? If he'd had time to do this right, to pack six suitcases, as he'd planned, to carefully choose his samples and then get on a plane out of Sana'a like the businessman he'd fashioned himself to be, he wouldn't be looking at a suitcase after midnight, having to choose which regions of Yemen would not be represented when he reintroduced the country's coffee to the world. He removed a dozen samples and closed the suitcase. He carried it downstairs and waited for his ride.

Sadeq pulled up in a sixteen-wheel flatbed. It was big enough to haul a car on top. And it was white. Any hope of slipping unseen through the night evaporated. They would be announcing themselves to anyone on the road and anyone bombing from above. A bright white flatbed truck moving through the Yemeni night in the middle of the most severe bombing campaign of the new war.

"Okay," Mokhtar said. "Let's go."

It was just after midnight. They had nine hours to get to Aden.

CHAPTER XXXI

THE ROAD TO ADEN

THEY LEFT THE CITY as it shook from another strike.

"We'll be okay," Sadeq said.

Mokhtar looked at this man. In the rush of preparations he hadn't considered that he didn't know these two people, Sadeq and Ahmed. He knew nothing about them other than that they were friends of Ali. They were both about his age. Sadeq had a tangle of wild black hair and wore a traditional outfit—more in keeping with the northern tribesmen than the urbane capital dwellers. Ahmed, with short hair and a well-kept beard, wore pants and a polo shirt. Mokhtar was about to drive nine hours with them, to meet a ship he knew nothing about. He didn't even know where the Greek ship was going.

They left the city without incident but knew that they'd soon be stopped at Houthi checkpoints. They were monitoring the movements of people and possible opposition, weapons, everything.

The two-lane road wound out of the city. They were going about 130 kilometers per hour, far too fast to be making the turns they were making. At the first checkpoint, Ahmed slowed as a trio of soldiers came into view. Mokhtar expected to be stopped, questioned, inspected. But the soldiers looked at the truck, at its front grille or license plate—Mokhtar couldn't tell—and waved them through.

The second checkpoint was different. The soldiers, in a mix of National army gear and Houthi garb, waved Ahmed to a stop.

"Where are you going?" they asked.

Ahmed told them the truth—that Mokhtar was trying to get out of the country through the port in Aden, that they were transporting a small sampling of Yemeni coffee. The soldiers wanted to see it. Mokhtar got out and untied the suitcase. He knew it looked unusual, and admitted to the soldiers that using a giant flatbed truck to carry one black suitcase had the appearance of something nefarious. He laughed. The Houthis did not share his mirth.

Mokhtar opened the case, showed the soldiers the beans, and soon heard himself explaining the history of coffee in Yemen, how he intended to restore the global stature of Yemeni beans. He went on and on as he always went on and on. The soldiers didn't care about the history of Yemeni coffee.

"Go," they said.

Ahmed drove away.

Every twenty minutes there was another checkpoint. Sometimes they stopped and explained their cargo and destination. Sometimes they opened the suitcase and revealed the coffee. Other times they were just waved through. Mokhtar couldn't get a grip on the system, if there was one. More and more it seemed odd when they weren't stopped. When this happened, he noticed a certain chain of events. Each time, the soldiers looked at the license plate, or something at the front of the truck, nodded, and let them pass. Mokhtar couldn't make sense of it, but they were making decent time and he had no reason to question it. They were on pace to make it to Aden by 8:00 a.m. He expected eventualities, unforeseen obstacles, but so far they were ahead of schedule.

At about one-thirty, they were stopped at a checkpoint. They explained where they were going.

"You headed through Yarim?" the soldier asked.

Ahmed and Mokhtar confirmed they were. Yarim was a small town between Sana'a and Ibb. It was barely on the map, but Mokhtar knew it well—he stopped there often when traveling north or south.

"There's going to be trouble down there," the soldier said.

"What do you mean?" Sadeq asked.

"Don't go through Yarim," the soldier said, and waved them on.

But they had no other options. The highway passed directly through Yarim. They shrugged it off. Maybe they were overconfident. The Houthi checkpoints had been easy. They felt invincible and had seen no evidence of the Saudi bombing since they'd left Sana'a.

They drove toward Yarim.

Yarim was twenty miles from the checkpoint where the soldier had given his cryptic warning, but five miles from the city, in the dark roadside, they began to see civilians fleeing, heading north, some walking, some running. It was 2:00 a.m.

"What the hell?" Ahmed said. Traffic slowed and soon stopped just outside of town. As they got closer to Yarim, the trickle of people fleeing turned into a flood. Thousands of people were fleeing along the roadside. Cars going north moved slowly. Cars going south, in the direction of Aden, weren't moving at all.

A man ran from the town screaming. He pointed at Sadeq's truck. "You're gonna burn! You're gonna burn!" he yelled.

Mokhtar got out. The air was unusually warm, and there was an acrid smell coming from the city—something was on fire. He asked another man, walking away from the town, what was happening.

"The Saudis just bombed the town," he said. "They were after an oil tanker. They hit a yogurt truck. They killed ten or more—kids, babies."

But it had happened only minutes earlier. How had that Houthi soldier known it was coming?

Ahead, the orange glow of multiple fires silhouetted the buildings in the town center.

"We have to go around," Sadeq said.

They turned off the main road and found a dirt road that circumvented the city. Their headlights illuminated the quick shadows of people running. Yarim was a cauldron.

"Do we stay? Help?" Mokhtar asked Sadeq and Ahmed.

Staying would serve no purpose. They couldn't help. They weren't firefighters or paramedics. And what about the chances their truck would be a target? The bombers had targeted oil tankers and had hit a yogurt truck. They needed to move on, get away from the city, from civilians.

But now the truck was stuck.

Mokhtar and Sadeq got out. They couldn't see the road or even the wheels. Mokhtar used his phone's flashlight to illuminate the problem. The wheels were spinning in six inches of mud. Pushing a six-ton truck was not an option. They tried wedging rocks under the tires, to no avail. They looked around for help, but the scene was chaos. No one would help a passing truck. In minutes the three of them were calf-deep in mud and the truck hadn't moved. They needed a tow.

Sadeq waved down a passing motorcycle and in seconds was on the back of it and was gone. He hadn't told Mokhtar or Ahmed what he was doing. Mokhtar stood in the dark. He stared into the black sky, the stars tiny and bright. They'd driven two hours from a city on fire into a town on fire.

Forty minutes later, a pair of headlights appeared. It was a truck. Sadeq jumped out. In the middle of Yarim, reeling from a Saudi bombing, he'd somehow found a tow truck. The driver dragged their rig from the mud. Ten minutes later they'd paid him and were on their way again. They sped around Yarim and were back on the highway. Mokhtar looked at his phone. It was almost 3:00 a.m.

His phone was running out of juice. They couldn't make it to Aden in time, he knew. And if Yarim was just bombed, other cities along the way could be bombed, too. Next time, they might not be so lucky as to arrive afterward. Now this plan seemed ill advised. In the truck, Mokhtar gave Sadeq and Ahmed the option to abandon the trip.

"We can turn around," he said. "This is all wrong. The Saudis are targeting trucks. That could have been us."

"But it wasn't," Sadeq said. "Think of it. God was looking after us." In the moment, the logic seemed sound.

They continued south to Aden.

CHAPTER XXXII

ADEN WELCOMES YOU

THE SUN WAS RISING when they entered Aden. As they
approached the center of the city, ten men waved the truck down and
surrounded it. Mokhtar assumed they were popular committee members. In 2011, popular committees arose in various parts of Yemen, to
defend territory against the Houthis and al-Qaeda. In general, popular committees were groups of local men who had banded together,
militia-style, as part-time soldiers in times of crisis.

These men were armed but dressed in civilian clothing. One man
pointed a gun into the cab. "Get out."

Mokhtar turned to Ahmed and Sadeq. "Let me talk to them."
They stepped down from the truck. The men converged on them,
frisking each of them. They searched the cab and found Mokhtar's
gun. News of the Colt .45 excited the crowd of men, and the gun
quickly disappeared.

Mokhtar told the group that he was trying to catch a ship from
Aden, that the two men with him were his drivers. He showed them
his passport, assuming that if they were popular committee members,

they would favor an American. The United States was ostensibly their ally.

But things were already out of his control. As he'd been talking, the men had heard Sadeq's accent and had seen his outfit.

"This guy's a Houthi," one of the militiamen said.

Mokhtar turned to find that Sadeq had already been blindfolded and there was a rifle pointed at his back. Mokhtar had to convince these men, who had been fighting the Houthis, who had already lost friends and family to the Houthis, that Sadeq was not a Houthi. But Sadeq seemed to have done everything he could to appear to be a Houthi.

"Itq'h allah, itq'h allah," Mokhtar told the men. This meant, "Fear God, fear God." It meant, "Slow down. Think of what you're doing. God's watching and will judge you for your actions."

"He's with me. He's just a driver. Not a Houthi," Mokhtar said, though he really didn't know who or what Sadeq was. Why had Sadeq agreed to make the drive in the first place? Didn't it make a certain kind of sense that he *was* a Houthi? Was he using Mokhtar to get into Aden?

"You're fine," one of the men said to Mokhtar. "You can go. But these two have to come with us."

Mokhtar knew he couldn't leave Ahmed and Sadeq. The men would kill Sadeq. Maybe Ahmed, too. Mokhtar told the militiamen that wherever they planned to take them, they'd have to take him, too.

"That's fine," they said.

Ahmed's blindfold was removed, and he was allowed to drive,

while a popular committee member sat in the cab with Mokhtar and Sadeq, directing Ahmed where to go.

They drove through Aden's narrow streets until they arrived at what seemed to be a school that had been converted into a popular committee base. Two dozen men milled about on the street outside. More could be seen in the windows. Most had AK-47s.

Mokhtar, Sadeq and Ahmed were ushered out of the truck. As they stepped onto the street, Mokhtar urged Sadeq to stay quiet. "Let me do the talking," he whispered.

They were led into a first-floor room. The room was largely bare but for a makeshift bed against the wall and a row of chairs facing it. They were told to sit on the bed, and were given cold water. Hospitality, even for prisoners, was always observed in Yemen.

The leader was a man in his forties, wearing a polo shirt, khaki pants and sandals. He asked Mokhtar for his passport. Sitting on the bed, Mokhtar complied. The leader stood in front of the row of chairs, where three other popular committee members sat. He flipped through the pages with great interest.

"When did you go to Dubai?" he asked.

Mokhtar wasn't sure if this was a test—the stamp indicated the date of entry, after all—or if the man simply didn't know how such stamps worked. Mokhtar remembered as best he could. "I was there for a specialty coffee event," he said.

"When did you go to Ethiopia?" the man asked.

Mokhtar tried to remember. It had been the year before. He guessed at the date, and the man moved on.

"When did you go to Paris?" he asked.

"I don't remember. March, I think," Mokhtar guessed.

"You went to a foreign country and you don't remember?" the man asked.

It struck Mokhtar that this man saw his passport as impossibly exotic. He saw Mokhtar as a Yemeni who was also American, who traveled freely to Ethiopia, Paris, Dubai, and yet couldn't remember the details.

"Listen," Mokhtar said. "I can't remember all the dates. I'm stressed." He told the man in the pink polo shirt that he was trying to catch a ship at the Red Sea port, that he hadn't been prepared to be quizzed about the dates stamped on his passport. "I just work in coffee," he said. "Look in my suitcase—it's just coffee samples. I'm trying to help people, help farmers." He was talking quickly, and felt some security in their attention, their patience. He needed to stay alive, to keep his companions alive. He needed to keep talking.

"I was just going about my business, like you all were, when these damned Houthis fucked everything up."

He could see his captors' expressions change. Their postures softened. Feeling an opening, Mokhtar did something that surprised himself. He got up from the bed and walked over to the other side of the room, where his inquisitors were sitting, and he sat with them, looking back to Ahmed and Sadeq.

"You're trying to defend your city, your homes," he continued, "and these goddamned Houthis are invading. They have no right." He kept talking while looking to Ahmed and Sadeq, as if he'd symbolically separated himself from them and joined the side of the popular committee.

"We shouldn't have come to Aden. I knew it was dangerous,"

Mokhtar said. "I just wanted to be with my fiancée, Summer. We just wanted to go home."

All the while, he maintained casual eye contact with Ahmed and Sadeq, to make sure they didn't interrupt or contradict him. His plan, so far, had succeeded. By explaining his business and inventing a fiancée, he'd humanized himself. Now he had to do the same for Ahmed and Sadeq.

"And these two guys were nice enough to help me. I know they look suspicious. This one looks like a hillbilly," he said, pointing to Sadeq. "But that's just because they're so oblivious. They're just guys who worked at the mill, my coffee mill, and agreed to take me to the port."

It was almost true.

"We're on your side, guys," Mokhtar added.

With that, the tension was gone. Mokhtar knew they would not die.

They were led outside, and it was only then that Mokhtar saw that on the front of the truck, the truck they'd driven nine hours through the night, there was a bumper sticker that said GOD IS GREAT. DEATH TO AMERICA. It was the Houthi slogan, rendered in Iranian colors. It was no wonder they'd been waved through all the Houthi checkpoints.

He didn't know if their captors had seen it, too. He had to assume they had. He had no choice but to point it out. He did, and laughed.

"See?" he said. "We didn't even know that was there. No wonder we made it through so many checkpoints so easily."

The popular committee members chuckled, but now there was a new uneasiness. He'd miscalculated. They hadn't seen the bumper sticker. Now there were too many suspicious elements to Mokhtar and his companions: Sadeq's clothing, Mokhtar's American passport and this big empty truck, the three of them driving through the night, entering Aden when everyone else was fleeing. Now this bumper sticker. Mokhtar knew he had to lay it on thick.

"May Allah bless you all," he said, moving toward the truck. "I wish victory for you guys. I really do. And I think you can do it! I bet you will. I bet you're gonna beat those Houthi scumbags."

He started shaking their hands, patting their backs. He smiled, laughed, made it seem that he was some kind of visiting American dignitary inspecting the troops. He thanked them a few more times, and somehow it worked. They were free. Twenty minutes earlier, ten AK-47s had been pointed at their chests and heads. Now they were friends and at liberty to go. The only catch was that the popular committee members had taken Mokhtar's Colt .45.

Let it go, he told himself.

They got into their truck, and at that moment a black SUV pulled into the roundabout. An older woman in a niqab jumped out and greeted Mokhtar in a thick Brooklyn accent. "There you are!" she roared.

This was Summer's mom. She was glad to see Mokhtar, to see he was safe. Her hands were all over the place, gesturing. Mokhtar assumed Summer was in the SUV, too, so he went to one of the two open windows at the back of the SUV and found a young woman, also in a niqab.

"I'm sorry about all this," he said.

"I'm not Summer," the woman said.

"I'm over here," another voice said.

Mokhtar went to the other side of the car, to another pair of eyes looking out of a niqab. "It's me," she said. As she'd been waiting for Mokhtar, she said, the Greek ship had left. She expected there to be other ships, other opportunities.

Mokhtar relayed a quick version of their captivity and release. In telling the story, he let them know about the Colt .45 the popular committee had taken, and this, more than any other part of the story, grabbed the attention of Summer's mother.

"They took your gun?" she said. She was outraged. Her tone was the same as it would have been if some boys on a Brooklyn playground had stolen her son's basketball. "You have to get it back," she said.

A cloud of dust approached. A white Montero pulled up and a man leapt out. "What's happening here?" Tall and well-dressed, he seemed to be a person of some significance.

Summer's mother took over. She told the man that Mokhtar had been robbed by the popular committee, that he'd been detained, that his handgun had been stolen. Mokhtar was a very important person in the United States, she said. He was a rich man and employed eighty thousand people. And he was her daughter's fiancée.

Mokhtar had no choice but to go along. He couldn't contradict her. He looked to Summer for help. Could she get her mother to dial it down? Summer's eyes said, *Stay out of her way*.

This new man's name was Mokhtar, too, and he shared Summer's mother's outrage. He promised he would straighten all this out. He would get Mokhtar's gun back.

A crack opened the sky.

"Snipers," the other Mokhtar said. He pointed to the rooftops around them. Somewhere amid the five- and six-story buildings, Houthi snipers were aiming at popular committee soldiers.

"Let's go somewhere safer," Other Mokhtar said. "I have a hotel. Follow me. We'll make calls from there. I'll get your gun."

Summer's car was going home, to her family's place in Aden. She'd call Mokhtar later, she said. Mokhtar apologized again, and the three cars left the roundabout, Mokhtar, Ahmed and Sadeq following Other Mokhtar. Ahmed and Sadeq looked at Mokhtar. They'd had an opportunity to leave but now were heading back into the center of Aden to retrieve a gun. This seemed unwise.

CHAPTER XXXIII

OTHER MOKHTAR

AS THEY FOLLOWED OTHER Mokhtar through Aden, Mokhtar began to have doubts about this plan, too. Did he really need the gun? His grandfather had given it to him, and for that reason alone it was worth retrieving. But then again, they were in the middle of a city about to be overtaken by the Houthis. How many hours did they have before the city was surrounded?

"We should just go back," Mokhtar said.

"Go back to Sana'a?" Sadeq asked. "Then we definitely need a gun."

Mokhtar thought about it. Twenty checkpoints heading north. They might need a gun. And Other Mokhtar had promised swift justice.

They followed Other Mokhtar to his hotel. The streets were empty as they entered the lobby. Inside, except for the four guards holding AK-47s, the hotel seemed welcoming, even luxurious. The lobby was wide and clean, and they stepped across the gleaming marble floors, feeling the strangeness of relative splendor amid what would be, surely and soon, a war zone. (The next day, in fact, the hotel would be hit by a mortar.)

Sadeq threw himself on one of the overstuffed black leather couches and directed his attention to an Egyptian movie on a big-screen television. Other Mokhtar disappeared behind the desk and returned with a key. "Go, get showered, relax," he said. In the meantime, he said, he'd make some calls and straighten out the business of the gun.

Mokhtar, Sadeq and Ahmed found themselves in an elevator, tinny music wafting from above, rising to room 303.

They opened the door to a clean room with two double beds.

"I'm taking a shower," Sadeq said. Ahmed took one afterward, as if they were all on vacation and getting ready for a night out. Mokhtar was too tense to shower, to change. He sat on the bed, looked out the window, paced the room. *Was the Greek ship actually gone?* he wondered. Summer had seemed unconcerned about that. There would be others, she had promised.

A knock at the door startled him. He opened it to find a waiter bringing mango juice and cookies.

"Compliments of Mokhtar," the waiter said.

It was the only sustenance they'd had since they left Sana'a eleven hours earlier. Something about the gesture was comforting, even lulling. After they devoured the snacks, Mokhtar found himself suddenly off guard and very tired. He knew it was ill advised to take a nap in circumstances like this, but he lay down on a bed and in seconds he was asleep.

He woke up forty-five minutes later. He asked Ahmed and Sadeq if they'd heard anything from Other Mokhtar. They hadn't. Refreshed from the nap, Mokhtar was determined to get out of Aden. We'll get the gun sent to me, he thought. He called Other Mokhtar and told him they planned to leave.

"No, no," Other Mokhtar said. "I'll get the gun. Give me an hour. Get some lunch and call me afterward."

It was difficult, weeks and months later, for Mokhtar to explain exactly why he thought it was a good idea to go looking for lunch in the middle of a city under attack. But they got in the truck and drove through the city, looking for a restaurant and, finding nothing, pulled over to ask a man on the sidewalk where they could get lunch.

Sadeq did the inquiring, which was a mistake, for the man heard Sadeq's accent and immediately changed his posture.

"You a Houthi?" he asked.

Mokhtar tried to rectify things, using his best classical Arabic. He convinced the man they were not members of the sleeper cells that had sent Aden into a paranoid abyss. They turned the truck around and returned to the hotel. He called Other Mokhtar again.

"You get the gun yet?" he asked.

Other Mokhtar had not. The situation was a mess, he said. The popular committees were nothing like an organized body with efficient information flow. He hadn't been able to get a clear answer from anyone.

"But I'll send it to you," he said. "I promise I'll get it and send it up to Sana'a."

Mokhtar said that sounded fine, now writing off the gun completely. He asked Other Mokhtar how they could best get out of the city. Would Other Mokhtar help, or at least guarantee safe passage? He mentioned the suspicious man on the sidewalk. And of course there had been the armed welcome they'd received in the first place.

"No problem," Other Mokhtar said. "You have my phone number. Just tell them you know me. You'll be fine."

CHAPTER XXXIV

A QUICK DEATH FROM
A CLEAN BLADE

THEY'D FAILED AT GETTING to the Greek ship in time—if there
had ever been a Greek ship. And now they had to turn around and
make the same nine-hour drive north, all while the Houthis were
heading south. But at least they had safe passage out of Aden. Other
Mokhtar had given them a specific route that he said would avoid
most of the checkpoints. They'd be on the highway in minutes.
Outside of Aden, the roads were controlled by the Houthis, and their
bumper sticker would take care of efficient passage all the way to
Sana'a.

They drove three blocks from the hotel and encountered their first
checkpoint. They stopped, Mokhtar leaned over to explain their situation, and they were waved through. *Easy,* he thought. He settled back
in his seat and thought of Ibb. They'd stop in Ibb and eat well and
rest. His aunt would make them a feast. Sadeq and Ahmed deserved
that much.

Ahmed stopped again. Another checkpoint. A crowd of men, all
with AK-47s, gathered around the truck.

"Where the fuck are you going?" one man asked.

Mokhtar did the talking. He explained the coffee, Summer, the Greek ship, Other Mokhtar.

The men were skeptical. Sadeq looked to Mokhtar, his eyes worried. Finally a man stepped through the crowd. He was wearing a tank top, long shorts and sandals. "They're okay," he said. He was in his late twenties, younger than many of the popular committee members around him, but he seemed to carry extraordinary weight among them. He looked directly at Mokhtar, and something like trust passed between them. "You have a kind face," Tank Top said. "You have the aura of a nice person. Let them pass."

The men stepped back from the truck, and Ahmed sped away.

There was no time to contemplate what just happened. There was no pattern to the checkpoints. What happened at each was always different and entirely subjective. A crowd of men could be convinced of anything—that Mokhtar and his companions were Houthis, that they were allies, that they were or weren't dangerous. Every situation was fluid, eminently volatile. Aden was under siege and there would be little regret for any man killed under those circumstances.

Ahmed drove on. They had only to make it to the coast and they were okay. Another checkpoint stopped them, but a quick conversation granted them passage. Another block, another checkpoint. A few times Mokhtar leaned forward to mention Other Mokhtar, but no one he spoke to seemed to know who this Other Mokhtar was. Still, they made it through five checkpoints and could glimpse the blue sea ahead.

"Almost there," Mokhtar said. The sky was wide open, the day sunny and clear. Mokhtar could picture the lamb, the beef, the spread his aunt would lay out for them. He'd show Sadeq and Ahmed the view from Hamood's sixth-floor balcony—you could see all of Ibb from there, a hundred miles in any direction. They'd stay there until they were rested. They'd sleep for days.

"Shit," Sadeq said. "Look."

Another checkpoint, this one on the coastal road at Al Bura'aiqah—a beautiful beach known for its powder-white sands. Ahmed slowed and stopped. Fifteen men surrounded the truck.

"Get out," one of them said. He was in his thirties, clean-shaven, wearing a windbreaker and track pants. Across his forehead he wore a thin black bandanna. Mokhtar couldn't help thinking of the Karate Kid.

Ahmed got out, his hands raised above his head. Sadeq followed. Mokhtar crawled through the truck cab and out into the bright sun.

Karate Kid asked where they were going, and Mokhtar explained. He showed him his American passport. Karate Kid seemed impressed. There was a discussion between him and the other men, and calmly Karate Kid returned and said, "Don't worry. You're in my command."

Mokhtar smiled to himself. Where did Karate Kid get a phrase like that? *You're in my command.* All these men playing soldier—they took themselves so seriously. Karate Kid put his hand on Mokhtar's shoulder.

"You're fine," he said, his face grave. "You're an American. You don't have to worry. But these two . . . I recommend you leave them."

"I can't," Mokhtar said.

"We have to take these two to the police station for some questions. You can stay here."

"Then you'll bring them back?"

253

"We will," Karate Kid said.

There was something in the orderly attitude of the men at this checkpoint, something in the gentle and businesslike demeanor of Karate Kid, that put Mokhtar, Sadeq and Ahmed at ease. Karate Kid calmly said they would be taking Sadeq and Ahmed to a police station, and his associates calmly blindfolded both men and then calmly put them in the back of a white Hilux. Mokhtar, hypnotized by the routine nature of it all, by the casual and efficient way they all went about their business and saw this as a simple bureaucratic procedure, nothing to worry about. He did as he was told, which was to relax, it would only take an hour, no problem.

Mokhtar stepped from the road to the white sands of the beach and sat down. A small group of popular committee members sat with him, and together they looked out to the Gulf of Aden. Three American warships stood in plain view, a few miles away. Mokhtar thought of how much easier it would have been had the American government simply evacuated its own citizens. *There are three ships right there. My God.*

Still, the beach was beautiful. It was empty but for him and his guardians. Mokhtar took off his shoes and buried his feet in the fine white sand, the texture of ash. He ran his hands through it, raising his face to the sun. The soldiers asked Mokhtar about his work, and he found himself showing them photos on his phone—of Haymah, of Bura'a, Hajjah, Bani Matar, Ibb, Utmah—all his pictures of the mountain terraces, the astonishments of high-altitude agriculture.

"That's *Yemen?*" one of the younger soldiers asked. He'd never left Aden. He had no idea there were landscapes like that in Yemen.

Mokhtar said yes, that's Yemen, that it was all Yemen, that there was so much more to the country than Aden, than Sana'a. More men gathered around Mokhtar as he waved his fingers over his screen, left and right, showing them the drying beds, the red cherries, the bright green leaves, the tanned faces of the farmers, their children.

Another man, younger than the last, asked the same thing: "That's really Yemen?"

The roar of an approaching vehicle snapped them out of their reverie. A white Hilux swerved across the highway and a large man jumped from the truck bed. Immediately he zeroed in on Mokhtar.

"Come here!" he yelled.

He was wearing a tracksuit and a leather jacket. Five other men, most of them armed, remained in the back of the truck.

The atmosphere changed so quickly and radically that Mokhtar found himself leaping to his feet and rushing to meet the man on the road, leaving his sandals on the beach.

"You're coming with us. In the truck," the man in the leather jacket said.

Mokhtar didn't see any room in the truck bed, so he moved toward the passenger seat of the cab. Everything had seemed so friendly until then that sitting in the passenger seat seemed most logical, most hospitable, among these new friends of his.

"No!" the man in leather yelled. "In the back!"

Mokhtar made his way to the truck bed. Another man grabbed him and began tying his hands behind his back. The material was soft—it seemed like a length of fabric torn from a shirt. Mokhtar wanted to get his sandals from the beach, but he knew it was too late. Whatever was about to happen, he would be barefoot for it.

Now he was being blindfolded. The blindfold was hastily applied, so Mokhtar could still see through the bottom of the fabric with his right eye—a glimpse of the ground in front of him.

He was helped into the back of the truck, where he sat on the wheel well, and in seconds they were on the road again. The wind rushed by, the air getting denser. They were heading into the city.

"You're a goddamn Houthi," the leather man yelled through the wind. He was sitting in the truck bed with him.

"I'm not a Houthi," Mokhtar said. He tried to remain calm, speaking in his best classical Arabic. He knew they would be listening for any trace of a northern accent.

"We plan to kill you," the man in leather said.

Again Mokhtar tried to remain placid and wise, to project the air of a neutral man, a bystander, a civilized citizen of the world caught in a fight not his own. "Do you really want that on your conscience?" he asked.

"I have plenty of dead men on my conscience," the man in leather said. "I killed two of you earlier today."

Now Mokhtar believed he might die. All the other popular committee members had been ordinary men and teenagers and middle-aged bank managers forced to take up arms to defend Aden. But this man was a thug, an opportunist, perhaps a madman. Whether or not he believed Mokhtar was a Houthi, he might actually kill him.

The truck made drastic turns through the city. Mokhtar, smelling diesel, feeling the shadows of tall buildings, contemplated how he might die. Would they shoot him? He thought of the burn of lead passing through his skull. He remembered his uncle Rafik, the Oakland cop, telling him that there was a spot between the eyebrows and the bridge of the nose—if you put a bullet there, he said, it was like an off switch. No pain. Just the end.

Mokhtar didn't want to be shot. A knife across the throat, he decided, would be his preference. The men in the truck all had machetes. He thought about asking the man in leather to grant him this—a quick death from a clean blade. That's how Muslims slaughtered animals to ensure meat was halal. It was a quick and humane death for mammals. He had a vision of his funeral. He pictured the mayor of San Francisco eulogizing him. President Obama might say something, he thought. At least relay a message. *Mokhtar Alkhanshali died doing what he loved doing.* Was that the right way to die? He thought of his parents, his siblings. He thought of Willem, Jodi, Marlee, Stephen. His death might be a source of inspiration. He would be a martyr. A coffee martyr? He'd lived well. For a few years, at least, he'd lived well. He thought of Treasure Island. Of the Infinity. The sculpture of the coffee-drinking man. No, his story wasn't such a good story. Not one that would mean much to anyone. A story without an ending.

He thought of his grandmother at her store in Richgrove. He saw Hamood and his aunt, preparing the great meal in Ibb. *Who would eat all that food?* he wondered.

He would die very young. He realized this with a shock. Twenty-five was very young. He thought of Miriam. Justin. Jeremy. Giuliano. They would live on, with the burden of their dead friend.

The truck sped through the city. Mokhtar figured it couldn't hurt to try something. He'd tell these men his story. He had nothing else.

"Do you want to hear my story?" he asked the man in leather.

The man in leather scoffed. "You sure you have it memorized?"

"*Inshallah,* I do," Mokhtar said.

The men in the truck laughed.

Mokhtar began, weaving a story of the coffee workers in Haymah and Bura'a, how he was organizing them, trying to improve their methods to prove that Yemeni coffee could be among the best in the world. The whole planet drinks coffee, but it was born here, he said. We should be proud of this. The world should know this. We have the chance to make coffee great, to show the world we have more than civil war and drones and qat.

When he finished, no one said anything. Mokhtar wasn't sure the man in leather, or anyone else, had heard him. And now the wind was howling, and they were thundering over a broken road, speeding to a sudden end.

CHAPTER XXXV

A GENTLE HAND

THE TRUCK STOPPED, AND tipped right and left as the men jumped off. Someone, a gentle hand, took him under the arm.

"Careful here," the man's voice said. It was a friendly voice, as gentle as the arm helping him. "Step down."

Mokhtar stepped onto the street.

"Just through here," the gentle voice said.

Mokhtar realized he wasn't dead and might not die. This new voice—who was this gentle new voice? Was it possible that this man would gently lead him to his death? Where was the man in leather?

"Up these steps," the gentle voice said. Mokhtar stepped up a flight of stairs and was led down a hallway and through a door. Even blindfolded, he could sense the change in light. He was in a dark space. He flinched, assaulted by rank smells. The air was dense and humid with human sweat, the odor of unwashed men, of stale urine and feces.

His blindfold was removed, and he took in his surroundings. He was in a small room full of filthy men. A steel-cage door closed behind him. It was a jail cell within a police station.

He saw Ahmed, who rushed to him.

"Where's Sadeq?" Mokhtar asked, but then saw that Sadeq was there, too, just behind Ahmed.

"I've been with you all along," Sadeq said. He'd been on the truck, too.

Mokhtar felt above his belt for his money. It was still there. In all this time no one had frisked him. If they had, surely they would have found it. He still had four thousand dollars strapped to his stomach.

There were ten other men in the room, all wearing rags. One man was asleep on the cement floor, covered in human waste. Men had been defecating everywhere. Mokhtar couldn't breathe. The smell choked him, his eyes watered.

"What is this place?" he whispered to Ahmed.

"Jail," Ahmed said.

There was one barred window high on the wall.

Most of the men appeared to be far gone. They might have been patients in a nightmare facility for the mentally ill. Mokhtar thought of the men he'd seen on the streets of the Tenderloin. In the corner, a man lifted his sarong and squatted. A pool of urine spread around him, a rivulet making its way toward Mokhtar's bare feet. He backed away, almost running into another prisoner. He asked him how long they'd been there. He expected the man to say months, given his ragged appearance, but the man said, "Four days."

Will that be us in four days? Mokhtar thought. *Had these been normal men four days ago? No,* he thought. Their clothes were scraps, they were muttering to themselves. The popular committees, he surmised, had moved them here to get them out of the way during the fighting. That was the only explanation. They'd rounded up the town's mentally ill—anyone usually on the street—and put them here for their own safety.

"I'm gonna fuck your grandmother!" a raspy voice yelled. It was an older man, barefoot and drooling, who continued to scream similar sentiments for the next hour.

Mokhtar found a place at the back wall and leaned against it. Ahmed came close. "I can't handle this," he said. His eyes were desperate.

"Relax," Mokhtar said. "There's nothing we can do. At least we're safe here."

But Ahmed wanted out.

"I'm gonna fuck your grandmother!" the old man yelled again.

Mokhtar moved closer to the cell door.

He overheard two guards talking about the port of Mokha. One guard had family in Djibouti, and apparently there were still ships moving between the two coasts, carrying onions and cattle from Mokha to Djibouti City.

"I'm gonna fuck your mother. And your sister!"

"Guard!" Ahmed roared. The crazed man was loud but Ahmed was louder. He was losing his grip. "Guard! Guard! Please! Please!"

One of the prisoners, this one relatively clean, approached Mokhtar.

"You met Ammar?" the man said. He described the man in leather. "You're lucky you lived. See that blood over there?" He pointed to a dark corner. "That's where he beheaded two guys."

Mokhtar didn't want to look. He wasn't sure he believed this man, who seemed the sanest madman in the cell.

"Guard!" Ahmed yelled again.

Mokhtar and Sadeq tried to calm Ahmed, to no avail. Ahmed couldn't hold on. "Guard! Guard! Guard!" he wailed.

"Calm down," Mokhtar said. "Go to the window. Get some air."

Ahmed continued yelling. "Guard! Guard!"

"Idiot! You'll get us killed!" Sadeq hissed.

Finally a young guard came to the door. Ahmed rushed to him, reaching between the bars to grip the guard's knees. He did his best to kiss them. In Yemen it was a traditional sign of supplication.

Another official of some kind, an older man with a long beard, appeared behind the guard. He wore a sarong and a polo shirt. He seemed to be a man of some local importance, the highest-ranking official they'd seen since being brought to the station.

Mokhtar went to him. A beard like that was almost surely a sign that the man was religious—that he might have an appreciation for the schooling Mokhtar had received as a young man.

"Sir," Mokhtar said. "I think there's been a mistake. We don't belong among these men. I'm a scholar. I spent a year in a madrassa." Mokhtar had hated his year at the madrassa, but he hoped now it would pay dividends.

"Where are you from?" the man asked.

"Ibb," Mokhtar said. "I studied with your scholars."

The man's attention was piqued. Mokhtar recited Quranic verses in classical Arabic. He chose verses relating to mercy, to hospitality, to the treatment of prisoners.

The bearded man turned to the guard. "These men don't belong here. Open the door."

Ahmed rose to his feet. Sadeq came to the door. The guard opened it, and the bearded man led them upstairs, one guard trailing them, until they came to a small office. There was a desk but little else.

The door opened and a familiar face appeared. It took Mokhtar a moment to remember where he'd seen this man. Then he remembered. It was the young man in a tank top and shorts who had told Mokhtar he liked his face. Now, seeing Mokhtar as a prisoner, Tank Top was furious.

"What's happening here?" he demanded.

The bearded man shrugged. The dynamic between the two men was hard to read. The bearded man was at least fifteen years older, but Tank Top seemed to outrank him. Tank Top roared and stomped around, apoplectic about the treatment of Mokhtar and his friends.

"I can't believe this!" he yelled. "What are these men doing here? Who approved this?"

Mokhtar watched Tank Top carefully. The theatricality of his outrage reminded Mokhtar of an amateur good cop/bad cop routine. Tank Top pounded the desk, pounded the wall.

Another official entered the room. He was clean-shaven and white haired. On a normal day he might have been the police chief. Now he was dressed in civilian clothes, like the rest of the popular committee. He assured Mokhtar, Ahmed and Sadeq that it was all a mistake. Ammar, the man in leather, was a rogue, he said, and they were free to go.

The bearded man stepped close to Mokhtar.

"My wife is from Ibb, too," he said. "I'm sorry about all this." His name was Abdul Wasr. He gave Mokhtar his phone number, promising to help if they needed it. Soon the group was laughing about the wild events of the day, the madness of it all. Mokhtar told them about his coffee work, and they told him how impressed they were. Then it was time to go.

And as with their earlier interrogation, Mokhtar walked out of his captivity in bright camaraderie with his former jailers. As they left the police station, Mokhtar was still telling them how much he could help their cause on the public relations front. He'd translate their messages into English, he said. He'd help them set up a Twitter account, get them on Facebook—he could do all their social media. He was from San Francisco and knew all kinds of people in Silicon Valley, he said. Meanwhile, Ahmed was kneeling before the police chief, kissing his knees.

CHAPTER XXXVI

SIX ARMED MEN AT THE FOOT OF THE BED

BUT THEIR TRUCK WAS still at the beach where they'd first been stopped. So Tank Top drove them back to the coast, where they found their truck untouched, the Samsonite still lashed to the truck bed.

"It's getting dark," Tank Top said. It would be dangerous being on the road anywhere near Aden in the dark, he said. He suggested they stay the night and leave in the morning. He knew a safe place.

Mokhtar had a feeling that this was a mistake, that they should get out while they could, but the afternoon had become evening and Tank Top, who had been so theatrical in his objections to their captivity, still had something in his eyes that indicated he didn't completely trust Mokhtar's trio. Leaving in a hurry would arouse more suspicion.

They got in their truck and followed Tank Top to the Al Ghadeer Hotel and parked in front, on a street otherwise devoid of life.

Tank Top got out of his SUV and led them into the lobby. He knew the proprietor, a thin, mustached man in his forties. "Take care of these guys," he said. "They're my friends."

The hotel was cheap and dingy. The three men said their goodbyes to Tank Top and made for the stairs.

"Oh, wait," Tank Top said. The three men turned around. "Be careful tonight. There are some guys around, guys like the one who picked you up earlier." Mokhtar knew he meant Ammar. "They go around at night making trouble. I don't think they'll come tonight, but just be aware."

Mokhtar, Sadeq and Ahmed were exhausted and emotionally wrecked. They collapsed on the beds, and Ahmed slowly recovered the composure he'd lost in the jail. Lying on the bed, staring at the wall, Mokhtar felt increasingly ill at ease. They shouldn't have stayed in Aden.

Sadeq, though, was acting like he was on vacation. He wanted to switch rooms to get a better view.

"Are you serious?" Mokhtar asked.

"This whole place is empty," Sadeq said. "I bet all the ocean-facing rooms are available. We could each get our own."

Mokhtar told him to forget it. Sulking, Sadeq went to take a shower.

"You okay?" Mokhtar asked Ahmed.

"I'm okay," Ahmed said. "But I was sure we'd die in that cell."

Sadeq emerged from the bathroom. He was wearing camouflage underwear.

"Are you crazy? Take those off!" Mokhtar yelled. It was unlikely that the popular committee would see Sadeq's underwear, but any camouflage clothing would imply their connection to the Houthis. They couldn't take chances. Sadeq took them off and Mokhtar hid them behind the dresser.

Now Mokhtar was confused about just who Sadeq was. Could he actually be some Houthi sleeper? He didn't want to ask.

Mokhtar turned on the television. The news showed footage of Houthis gaining ground all over the country. They were gathering a few miles from Aden. Ahmed sighed loudly. As gunfire rattled from the TV, Mokhtar fell asleep.

He opened his eyes and saw a row of shadows. It was 2:00 a.m and there were six new men in the room. All hid their faces under keffiyehs. All carried AK-47s, their fingers on the triggers.

Mokhtar could tell most of them were young. Before thinking, he said, *"Masa al-khair!"* (Good evening!) He said it as if he were hosting a dinner party. Immediately the sense of menace was punctured. The man directly in front of him seemed to be smiling. Mokhtar could see his creased eyes above his scarf.

"What do you mean, *Good evening?*" another man said. He seemed to be the leader.

Again Mokhtar spoke before he thought. "What do you want me to say, Good morning?"

Ahmed and Sadeq looked at Mokhtar with a mixture of horror and awe. His mouth was going to get them all killed. But Mokhtar felt confident that whatever he was doing was working. He'd talked them out of a few situations before, and this was no different. Already he'd made one or two of the masked men smile. The AKs were a bad sign, but other indicators were promising.

The scarves were a good sign. The scarves meant the men assumed that some or all of their captives—Mokhtar and his friends—were

Houthis, and that if they showed their faces, and later freed their prisoners, the Houthis might retaliate against them or their families. The real danger, he knew, was when a group like this didn't care if you saw their faces. Then you were dead.

The leader asked what they were doing in Aden. Mokhtar told them that they were trying to get to the port, where they would catch a Greek ship headed out. The Arabic word for "Greek" is *Yunani,* and speaking quickly, Mokhtar made it sound like *Irani*—implying they were looking for an Iranian ship. The masked men tensed. Iran was backing the Houthi insurgency.

"A *what* ship?" the leader said.

"*Yunani, Yunani,*" Mokhtar said. Greek, Greek.

He explained that he was an American, a coffee trader, a businessman, just trying to get out of Yemen with his bean samples.

"You're *proud* to be an American?" the leader asked.

Now Mokhtar got worried. Who *were* these guys? They seemed like popular committee, but they could be al-Qaeda. In Yemen there were strange bedfellows, including some unsettling overlap between the popular committees and AQAP.

"Give me your laptop," the leader said.

It was a terrible idea to surrender his laptop. In a flash, he remembered Michael Li from his days selling Hondas. *Control the conversation.*

"Okay, you can borrow it," Mokhtar said, "but I need it back by seven in the morning."

He said these words to a group of masked men carrying AK-47s. He insisted on getting his laptop back in five hours, as if he had a conference call at seven he couldn't miss. Stranger still, the man agreed.

"Okay," the leader said. "And we need your phones, too."

Mokhtar, Ahmed and Sadeq handed over their phones. The men took them and left the room.

Ahmed got up and locked the door. "What the hell was that?" he asked. "You told them you need your laptop back by *seven*? You want us killed?"

In the dark, Ahmed, Sadeq and Mokhtar speculated about who the men were. They contemplated an escape, but there was only one exit, guarded by either this team or the original popular committee.

"Let's sleep," Mokhtar said. He was tired and had the strange sense that he would sleep, and that at seven there would be a knock on the door, with one of the gunmen returning his laptop.

Instead they came at 5:00 a.m. Mokhtar felt someone nudge him on the shoulder. He opened his eyes and saw his laptop being returned by one of the masked men. He told the man to put it on the floor. The man obeyed, and Mokhtar went back to sleep.

In the morning Mokhtar woke with the rising sun and opened his laptop. It worked, and all seemed unchanged, except the background screen had been altered. He'd had an image of a mountain village in Haraz; now it was a picture of one of Mokhtar's flyers for a coffee event he'd held in Oakland. His captors didn't seem to have touched anything else. The only explanation he could conjure was that the popular committee men didn't know how to use Macs. No one he'd ever met in Yemen could use a Mac. These men had tinkered with the computer long enough to do only one thing: change the background screen.

He went downstairs. The lobby was dark. The exit was blocked by a roll-down door. They were locked in. The man behind the front desk was asleep in his chair.

From the front-desk phone, Mokhtar called Abdul Wasr, the bearded man who had helped free them from the police station. Mokhtar assumed he knew the situation.

"We need to leave," Mokhtar said. "Where are our phones?"

Abdul sounded troubled. "Are you alone?" he asked.

Mokhtar confirmed that he was.

"There are concerns about your friend Sadeq. They found some troubling names on his phone. The men wanted to take you all last night, but I convinced them not to."

Abdul told Mokhtar to stay at the hotel. He arrived a few minutes later, and informed Mokhtar that his own phone would be returned soon, but on Sadeq's phone they'd found numbers of high-ranking Houthis and Houthi collaborators in the Yemeni army. The popular committee planned to take Sadeq for further questioning. Mokhtar assumed he would be tortured.

Mokhtar went back to the room and confronted Sadeq.

Sadeq was blasé. "It's my cousin's phone. I don't know any Houthi generals."

Mokhtar asked why there were names of Houthi and Yemeni army generals in his cousin's phone. Sadeq said that his cousin had a delivery-service truck and a wide range of clients—hotels and schools, some military bases.

Mokhtar believed him. Sadeq was not some revolutionary. They'd entered a hotel in a war zone, and he'd asked for an ocean view.

Mokhtar went downstairs and explained this to Abdul. "It's not

270

possible that Sadeq is some operative for the Houthi forces. You saw him, right?"

Abdul allowed Mokhtar to use his phone. Mokhtar called Ali back in Sana'a.

"You're alive," Ali said. "Good. Andrew's already talking to the girl. To Summer."

"Summer?"

Ali explained, and Mokhtar put it together. The popular committee men had taken Mokhtar's phone last night, and as they were going through the phones, Andrew called Mokhtar. Not knowing how to use an iPhone, the popular committee men had tried to stop the ringing, but had inadvertently answered the phone and left it connected.

Andrew was able to hear about two hours' worth of conversations between the men of the popular committee. What he'd heard was disturbing; Mokhtar and his friends were in grave danger. Andrew and Ali spent the night making phone calls. They found Summer's name, called her, and she made calls until she found where Mokhtar was being held and by whom.

"Now it'll be okay," Ali told Mokhtar. "We have all kinds of people talking now. We'll make sure you're safe. Summer's family knows everyone in Aden."

Buoyed by new hope, Mokhtar went back up to the room. He showered, trying to wash the dirt from his feet—he'd been barefoot in the filthy cell the day before. That seemed so long ago now. Alone for the first time in a day as the water fell around him, he thought through the options. Sadeq was now a suspected Houthi operative who had infiltrated Aden, presumably to report enemy positions and capabilities back to his superiors. The popular committees defending Aden were justifiably paranoid about anyone within the city—especially a Houthi who had seen one of their strongholds in the police station.

Mokhtar didn't know what to believe or do. If Sadeq was a Houthi, could Mokhtar continue to defend him? Was it time to distance himself from Sadeq? And what about Ahmed? Was he complicit?

Sadeq had taken the only towel, so Mokhtar used the linen shower curtain to dry himself. When he walked out of the bathroom, he saw Other Mokhtar. He was with six armed men.

"We're getting you guys out of here," he said.

Other Mokhtar had been called by Summer.

"You have to leave the city now," he said.

Why this man was risking so much for Mokhtar, Ahmed and Sadeq was unclear. But Mokhtar couldn't question it. They were almost free. Then Mokhtar's mouth opened and said something that in the moment was absurd to all in the room.

"My samples. They're in a black Samsonite. I can't leave without them."

Other Mokhtar winced. "Your what?"

Mokhtar told the story of his coffee beans, the conference in Seattle.

"It's my whole life," Mokhtar said.

"Stay here," Other Mokhtar said.

He and the six armed men left.

"Are you serious?" Ahmed said. "We have a chance to leave now, and we're staying for your beans?"

An hour went by. Two hours.

Mokhtar, Sadeq and Ahmed watched the war on television. None of them knew the geography of Aden well, but it looked from the reports as if the fighting was all around them. Mentions of Houthis on the news gave Mokhtar a thought.

He looked at Sadeq. "We have to fix you."

Mokhtar had an extra dress shirt and gave it to Sadeq. Sadeq put it on. The transformation was profound but not finished. Mokhtar gave him his glasses and combed his hair so it looked orderly.

"That's incredible," Ahmed said, looking at Mokhtar. "He looks like you." Sadeq looked like a global businessman—tidy blue dress shirt, spectacles, hair parted neatly on the side. Instantly Mokhtar thought any threat had disappeared. Why hadn't he done this in Sana'a, before the journey? Too many of his best ideas occurred far after they would have been most useful.

A knock thundered through the room.

It was Other Mokhtar. "I've got your suitcase. Let's go."

Outside, Mokhtar saw the truck, the suitcase in the back.

"I have to stay with my hotel," Other Mokhtar said, "but my friend here will go along with you." He introduced them to a man named Ramsi. "If anyone asks who you are," Other Mokhtar said, "you're part of a water treatment company in Aden, and you're leaving."

Ahmed started the truck. Mokhtar thanked Other Mokhtar. He

owed his life to this man he would never see again. Ahmed pulled away.

There were three checkpoints before they were clear of Aden, and Ramsi talked them through each. They dropped off Ramsi ten miles out of Aden. After that, the roads were clear. They sped the nine hours back to Sana'a unimpeded and arrived that night.

CHAPTER XXXVII

THE PORT OF MOKHA

SITTING IN MOHAMED AND Kenza's apartment, Mokhtar weighed his options. In the popular committee prison, he'd heard a guard talking about freighters shipping livestock and people between Mokha and Djibouti City. Online, Mokhtar found that the port of Mokha was more or less functioning. It had been bombed by the Saudis repeatedly, and when it wasn't being bombed, it was being fought over by the Houthi and government forces—but ships were leaving regularly.

He called Andrew.

"You want to take a boat from *Mokha?*" he asked.

"We get to Djibouti and fly to Addis," Mokhtar said.

This time Andrew agreed. The Aden trip hadn't appealed to him, because Aden was an active war zone, and because he'd been holding out hope that a more practical solution would present itself—that the airport might reopen, for example. But that hadn't happened, and now the SCAA conference was fast approaching. Andrew counted on the conference for a good portion of Rayyan's yearly sales. He had to be there.

Mokhtar called the U.S. embassy in Djibouti, expecting nothing, but reached a human. He asked, hypothetically, if he and another American were to get passage across the Red Sea, and were able to make it to Djibouti by boat, would they be received by the U.S. embassy and helped in their return to America?

The embassy representative, a friendly woman whose pragmatism was emboldening, confirmed they would.

"We won't be put in some refugee camp?" Mokhtar asked.

"No, no," the woman said. "If you make it here, we'll help you in any way we can."

Mokhtar and Andrew decided they'd go on Friday, after *jumma*. Violence was less likely on the Islamic holy day, they assumed.

Ahmed agreed to go again—just two days after he had narrowly escaped Aden with his life. Mokhtar was humbled. He barely knew Ahmed a week ago, and now he was risking his life, again, for what—for Mokhtar and Andrew and coffee?

"We'll be okay," Mokhtar told him.

Through friends in Sana'a, he'd been connected with a man named Mahmoud, who knew the movements of ships leaving Mokha. Mahmoud assured Mokhtar that he'd take care of the details in getting them on the boat and out of Mokha.

"No problem," Mahmoud said.

In the morning Ahmed arrived at Mokhtar's door, driving a pickup. Mokhtar threw his suitcases in the truck bed. They drove across the city to pick up Ali and Andrew. Andrew came downstairs dressed for Friday prayers, wearing cologne and carrying five suitcases full of coffee beans and a basket of Jennifer's muffins. They took off, and Andrew showed Mokhtar and Ahmed a video on his phone of his daughter Rayyan—she'd been named after his coffee mill. She was two years old and, in the video, was talking about strawberries.

"Why'd you have to do that?" Mokhtar said. He didn't want to be thinking about Andrew's daughter when they were driving to Mokha. He wanted to be thinking about prosaic things. Seattle. Beans.

Goodbye Sana'a, Mokhtar thought. He was sure it would be there when he returned—he had no idea when—but there was also the possibility it would be radically altered again. There was no guarantee what the Saudis would do, what the Houthis would do. Yemen could become Syria.

They drove west, through the Haraz mountains. The road was narrow and winding, taking them as high as three thousand meters above sea level. The checkpoints came every ten or twenty miles, but with Ahmed doing the talking, the Houthis allowed them speedy passage.

They made it to Hodaidah and joined the north-south highway. They'd encountered no resistance in the four hours they'd been on the road. The highway was twenty miles inland from the coast and traversed a high flat plain. There were four lanes most of the way, and the checkpoints were infrequent and efficient. They arrived in Mokha by early evening.

Mokhtar had read about Mokha, and had named his company after Mokha, and had for years been enthralled with its history. But this was his first time seeing it. The road into the town was potholed and surrounded by crumbling stone dwellings, many abandoned. The fabled port had once been one of the most important in the world, but all that remained were some fifteen thousand impoverished souls. The city had fallen on hard times.

There was one functioning hotel in the city. When Mokhtar, Andrew and Ali walked in, they found a chaotic scene. Everyone who wanted to get out of Yemen through Mokha was there—Ethiopians, Eritreans, Somalis. At the front desk, the clerk was charging about five times what a room would have cost on a normal day. But they had no choice. They paid their money and went to their room.

Mokhtar called Mahmoud, who said he could arrange passage on a ship the next day. He arrived at the hotel an hour later and confirmed they could take a Somali cargo ship that usually brought livestock to Mokha, but had recently been converted to take humans out of Yemen. The next day, he said—or maybe Sunday, he amended—the ship would take 150 people, and a few tons of onions. He'd make sure they had room for Mokhtar and Andrew. The trip to Djibouti City would take anywhere from fifteen to twenty hours.

Andrew was worried. There was no guarantee of leaving the next day. And no certainty about when they would arrive. The math was bad for their roasting schedule. If they didn't leave the next day, on Saturday, they wouldn't get out of Djibouti the next day, and that meant they wouldn't make the 10:00 p.m. flight out of Addis Ababa, and that meant they wouldn't get back to the United States in time to roast their coffee. Their coffee had to be roasted and rested to be good, and if it wasn't good, there was no point in any of this.

"So we leave tomorrow," Mokhtar said.

They ate dinner in the hotel and turned in early. Chewing qat in their room, they heard the roar of three diesel buses pulling up. Through the window they saw dozens of Somalis disembark. Mokhtar assumed they would be on their ship the next day, too. Everyone in the city appeared desperate to leave.

"We could see the al-Shadhili Mosque tomorrow," Mokhtar suggested. He'd been thinking about it all day. It was the spiritual home of the original Monk of Mokha, Shaykh Ali Ibn Omar Alqurashi al-Shadhili—the man who first brewed coffee, who built the coffee trade.

Andrew looked at Mokhtar like he'd lost his mind. "We're not going to a mosque tomorrow," he said. "We're not on vacation. We're getting of here."

They woke at sunrise and contacted Mahmoud.

"There's a problem," he said, and Mokhtar knew the rest.

Nothing was simple in Yemen. If someone said they could get you on a boat, that was just the beginning of the conversation. It was never as easy as buying a ticket and getting on that boat. Mahmoud was saying there was no fuel and the ship wasn't leaving that day.

"When is it leaving?" Mokhtar asked.

"Hard to say," Mahmoud said.

Mokhtar asked Mahmoud about other possibilities. Mahmoud mentioned the outside chance of hiring what he called a viper boat. On it, the trip would take five to six hours to Djibouti, he said. Mokhtar pictured a speedboat, the sort favored by Caribbean drug dealers.

"I'll look into it," Mahmoud said.

Mokhtar knew what that meant. He had time.

His guide was a local judge and historian, Adel Fadh. Short and middle-aged, with a gentle demeanor, he led Mokhtar into the mosque, a humble structure undergoing significant repairs. They walked under scaffolding as morning light streamed through the high windows. Built to honor Shaykh Ali Ibn Omar Alqurashi al-Shadhili, the mosque retained a vibrating spirituality. Al-Shadhili, a Sufi monk, had gone to Harar, married an Ethiopian woman and brought the coffee plant—which hadn't been cultivated yet; it was still wild—back to Yemen. Here, in Mokha, he invented the dark brew now known as coffee. Local lore had it that it was al-Shadhili who was responsible for Mokha's ascendance to the center of the coffee trade. It was he who introduced coffee to traders who came to Mokha, and who extolled its medicinal qualities.

The mosque was over five hundred years old, and had been repaired many times, Adel Fadh explained. But there was so little money to keep it up now. With Mokha so poor, and the country at war, he feared for the future of the mosque and the town.

"We can restore this port to greatness," Mokhtar said. If he could get out of Yemen alive, and come back someday, he would see to it, he said. He had no idea how he would do it but he felt obligated to give the judge some semblance of hope.

Adel, a guileless man, listened intently, and Mokhtar realized that all the workers in the mosque were listening, too. He spoke about the modern coffee trade, the rise of specialty coffee, the imminent supremacy of Yemeni coffee, how Mokha could thrive again.

Mokhtar's phone rang. It was Mahmoud. He'd found a boat.

At the hotel, it turned out Mahmoud hadn't found a boat. They drove around the town, asking anyone they met about the possibility of renting a fishing vessel, a Zodiac, anything.

Finally Mahmoud called. He'd found a boat and captain who

could make the trip. Ali drove them to the shore. The pilot was a young man, about thirty, and the boat itself was tiny, about fourteen feet long, just a flat-hulled skiff—this was no Viper boat. It turned out Mahmoud had been trying to say *fiber* boat, not *viper* boat. Their escape vessel was a sorry thing, low and narrow, with a single sorry Yamaha outboard. It looked like it could be capsized by a tuna.

"We'll get soaked in that thing," Andrew noted.

They got back into the truck, looking for tarpaulins. They'd have to wrap the suitcases in the tarps and set them on the bottom of the boat to keep the coffee dry. It was agreed that Ahmed would stay behind and go to the Port Authority to get Mokhtar's and Andrew's passports stamped.

They split up. Ali drove them back into town, where Mokhtar and Andrew found a shop that sold tarps. They bought three and turned back to the beach. The truck, though, was low on gas, so they stopped to fill the tank. Sitting in the truck, they heard the rattle of gunfire coming from the waterfront, less than a mile away.

"Call Ahmed," Mokhtar said.

Andrew called. Ahmed's phone rang inside the truck. He'd left it behind. Mahmoud's Ford Taurus appeared. He tore into the parking lot and jumped out. There'd been a gunfight at the Port Authority, he said. He and Ahmed had been getting the passports stamped, when the Houthis started shooting at the Port Authority security officers. Or the security officers started shooting at the Houthis. It was chaos. Mahmoud and Ahmed had gotten separated.

Mokhtar and Andrew were paralyzed. They needed to find Ahmed, but going to the waterfront seemed suicidal. And he wouldn't be there anymore anyway. He would have fled.

"The hotel," Mokhtar said.

They were silent in the truck as Ali sped through the wide streets of Mokha. Mokhtar had the distinct feeling that Ahmed was dead. He couldn't have gotten lucky again. He'd made it through all that mess in Aden but this was one crisis too many.

Andrew saw Mokhtar's face. "Don't worry," he said. "He's fine."

They pulled into the hotel and jumped out.

Ahmed was standing in the lobby, unscathed.

"Hey," he said.

Mokhtar threw his arms around him.

Ahmed laughed. "I'm fine. It was nothing."

Mokhtar pulled back and looked at him. Minutes before, he'd been sure that Ahmed was gone. Now Ahmed was alive, and he had their passports. When the shooting began, he'd hidden the passports in his shirt and had slipped out of the building and into the parking lot, where he ran through the crossfire until he saw a passing motorcycle. He flagged it down, hopped on and directed the driver to the hotel.

Now he presented the passports to Mokhtar and Andrew as if he'd just performed some rudimentary act of processing. He'd had them stamped before the fighting broke out.

"You better go now," he said.

They got back into Ali's truck and made their way to another part of the shore. The skiff they'd hired was far from the fighting. At the beach, they were met by a pair of local police officers whose allegiance—to Houthis or the government—was unclear. Mokhtar pressed a bribe into their hands, and they were free to leave Mokha.

* * *

When they looked closer at the hired skiff, Andrew and Mokhtar laughed. Andrew had grown up on a lake in Louisiana, and this watercraft was smaller than the boats he'd used to go fishing. Could it really make it across the Red Sea? The man they'd hired to steer it seemed confident enough. He said he'd done it many times.

There was no extra motor. There was one paddle. There were no life vests. They had no idea if there were Saudi ships out there. Or if Saudi planes would attack a craft leaving the port. Or if the American navy was out there and might assume they were terrorists and blow them out of the water. There was also the possibility—probably greater than any other—that the captain would sell them to Somali pirates.

"Time to leave," Mokhtar said.

They rolled the suitcases in the tarps and set them on the floor of the boat. Between them, they had one hundred kilos of green beans. While the captain was prepping the engine, Mokhtar, Andrew, Ali and Ahmed made a plan for any eventualities.

Mokhtar and Andrew would call Ali and Ahmed when they got to the port of Djibouti, or within twenty-four hours. Before that time, Ali and Ahmed were to stay in Mokha. If Mokhtar and Andrew didn't call within that time, that meant something had gone wrong, that they'd likely been sold to pirates. In that case, Ali and Ahmed were authorized to kidnap relatives of the captain. It was the Yemeni way.

On the beach, all of this was discussed with a mixture of seriousness and dark humor. Their suitcases lined the boat floor, and everything was ready. All this time, though, as they prepped the boat and discussed eventualities, two small local kids, a boy and a girl, had been hovering. This wasn't unusual in itself—there were always local kids who took an interest in any vessel leaving the shore—but now these two kids jumped into the boat.

"Who are these kids?" Mokhtar asked the captain.

They were the children of a friend of his, the captain said. He was delivering them to their father in Djibouti. Mokhtar and Andrew briefly debated whether the presence of two children made the trip more perilous or less so.

"Let's go," Andrew said.

They said goodbye to Ali and Ahmed. Ali, who had reassured Andrew and Mokhtar with his explanations of collateral and possible retribution for anything that might befall them, now seemed strangely unsure.

"So you're really going?" he asked.

"We have to get to Seattle," Andrew said.

"Call from the boat," Ali said.

They helped the captain push the skiff into shallow water. The captain got in and took his position at the outboard motor.

"You know what? I've never been in a boat," Mokhtar said.

"You've never been in a boat like this?" Andrew asked.

"Never been in *any* boat," Mokhtar said.

Mokhtar had grown up in San Francisco, surrounded by water—oceans and bays and rivers, estuaries and lakes. He'd spent years in Yemen, a country with a twelve-hundred-mile coast. He'd gone to middle school on Treasure Island, an actual island. But he'd never been on a boat. He'd always wanted to, but the ferries and yachts and sailboats he'd seen throughout his youth seemed part of some unattainable other world.

His first experience with any watercraft was going to be in a tiny skiff leaving Yemen in the middle of a civil war.

He stepped in and they left the shore. They were carrying the first coffee to leave the port of Mokha in eighty years.

CHAPTER XXXVIII

DJIBOUTI WELCOMES YOU

THE WAVES SENT THEM to the floor. They recovered and laughed and braced themselves. The boat, so small and stiff, made every ripple calamitous. They were soaked in minutes. The children were soaked, too. They huddled together in the middle of the boat and said nothing for three hours.

As the shore disappeared behind them, Andrew took a bag of qat out of his backpack.

"Really?" Mokhtar said.

Andrew smiled. Qat would calm them. Qat would make the ride seem routine. They chewed, and the qat brought them to a state of contentment and philosophy, even while the water and wind demanded they yell over the noise. After the first hour, the sea calmed, and the qat kicked in. And because the sun was shining and their trust in the captain was growing with every passing mile, they relaxed into complacency. Mokhtar and Andrew, huddled in the middle of the boat, sitting on their samples, found themselves engaged in a ludicrous string of philosophical conversations, the kinds of things people talk about only while on qat, and maybe only while on the sea, on qat, between a war zone and an unknown coast.

They talked about God, and Mokhtar heard himself say, "If you believe there's only one path to God, then you're limiting God," and thought he was saying something so profound it might forever change their lives. They talked about practical matters, too—about their coffee work, their farms and farmers and plans. Because they felt they were making it safely across the Red Sea, and they knew that hundreds of others wanted to leave Yemen but couldn't, they conjured a plan where, if they made it to Djibouti safely, they would charter a boat that could hold 250 or so, and they'd ferry Yemenis and Americans and anyone else across the Red Sea, doing the work the U.S. State Department couldn't or wouldn't do. They'd call it Operation Arabian Mokha. They were utterly certain this would happen.

They made their way south by southwest until they hit the Bab el Mandeb, a narrow strait where the sea bottlenecked between northern Djibouti and southern Yemen. The gulf grew choppy and the wind picked up. They spent an hour wondering how wet they could get and how much water the boat could take, how likely it was their samples would take in at least some of the sea.

But soon the Djiboutian coast came into view, desolate and gray. They hugged the coast for the next few hours as night came on. They passed the occasional fisherman, saw the distant lights of aircraft overhead. Night claimed the sky, and they motored through black water. The first Djiboutian port they encountered was Obock, a tiny town on the easternmost point of the coast, and by no means the place where they intended to disembark. There was no U.S. embassy there, and few if any services. They had no desire to stop.

But the captain was stopping.

"Just a minute," the captain said.

He had to drop the kids off. Obock was an entry point for Yemeni refugees, he said. There was a United Nations refugee camp nearby. He intended to deposit the kids there, and they'd be on their way.

But this sounded wrong. Mokhtar's Tenderloin senses were now on high alert. They'd spent five hours growing increasingly complacent about the captain and now there was this. This, a sudden unannounced change of plans, was exactly the kind of thing they'd feared. But the captain's behavior was so nonchalant that they found themselves allowing him to approach the docks. Mokhtar hoped for a five-minute turnaround: the boy and girl would climb off, the boat would shove off.

He didn't expect to see anyone in uniform. But now there were two men on the dock, and the captain was throwing them a line.

"What are we doing?" Mokhtar asked the captain.

"Just dropping off the kids," he said. "Don't worry."

Now the men in uniform beckoned Mokhtar and Andrew to disembark.

"Who are these guys?" Andrew muttered.

Mokhtar had no idea. Coast Guard? Local police? They were wearing blue camouflage and carrying German G3 rifles.

"Where are you coming from?" one of the officers asked.

"Yemen," the captain said.

"And these two?" the officer asked, indicating Mokhtar and Andrew.

The captain told them they were Americans stopping briefly on their way to the capital, where they were expected by the embassy.

Now the officers were very interested.

"You're different than the ones that came before," one of them said.

"What do you mean?" Mokhtar asked.

"The Americans that came before you. They had people from the U.S. government pick them up. Why don't you have anyone here receiving you?"

"Because we're headed for Djibouti City," Mokhtar said. "We had no plan to stop in Obock."

"Come with us," one of the guards said. "The governor will want to meet you."

Grim possibilities ran through Mokhtar's mind. Secret prisons. Illegal detentions. This could be a CIA black site. Since 9/11, Djibouti had been a significant U.S. counterterrorism partner. It was the launch point for the drone fleet that routinely bombed Yemen, and terrorism suspects had been detained, interrogated and tortured in Djibouti for years. He was thankful for the presence of Andrew, a white American. They wouldn't disappear a man like Andrew.

The captain had turned off the engine, and the kids were on the dock, and the Djiboutian men in uniform were welcoming Mokhtar and Andrew with wide smiles.

Mokhtar looked to Andrew. Nothing good could come of this. But because refusing seemed more fraught than accepting the governor's hospitality, Mokhtar and Andrew allowed themselves to be helped out of the boat, and soon their suitcases were out of the boat, too, and were being opened and inspected.

Suitcases full of small plastic bags. It looked like drugs. Mokhtar and Andrew had to explain their samples, their work in coffee, their schedule, their need to get back in the boat and get going, the conference in Seattle.

But getting there on time now seemed unlikely. They were being detained. Not in a hostile way. Not in a way that felt especially menacing—not yet at least. It was more like the disorganized and irrational detentions common at American airports, the kind of detention that came from the officers feeling they'd been confronted with something beyond their immediate comprehension, something too unusual to simply allow.

Mokhtar and Andrew found themselves in the back of an SUV, being driven to what they had been told would be the governor's home.

"You think that's where we're going?" Andrew whispered.

"I don't know," Mokhtar said while thinking it was brilliant, really, for the Djiboutians to tell them they were being taken to a governor's house. The prospect of being honored this way was meant to make them complacent. And where was the captain of their boat now? He was gone. The children were gone. This heightened the likelihood that the captain had somehow arranged their sale or transfer. The kids were a decoy! Mokhtar's mind was a jumble of dark-hearted possibilities.

The SUV pulled up in front of what looked to be a house. Not a prison. The officers opened the car doors, and Mokhtar and Andrew were led to the front door, where a smiling Djiboutian man wearing khakis and a button-down shirt walked out and greeted them.

"Hello, hello," he said, then turned to an aide. "Water? Can we get you some water?" he asked, leading them inside.

They accepted warm bottled water. It was easily 110 degrees in Obock, and the humidity was stifling. The governor led them to his office, a spacious wood-paneled room with a view of the sea.

He asked about their trip and their plans, and Mokhtar and Andrew told him how they'd hoped to go straight to the capital, and from there catch the earliest flight to Addis.

"Oh, you won't make that flight!" the governor said cheerfully.

It was already eight o'clock, and by boat the trip was a few hours, he said. And besides, they had much clarifying to do. The governor would have to talk to the U.S. embassy in the capital and let them know about all this, the unannounced arrival of two Americans in Yemeni dress.

"Spend the night here," the governor said. "We have a very nice tourist hotel. You'll stay."

They were not being given a choice. For the night, at least, they would be the paying prisoners of the local Djiboutian authorities. This was the second time in a week that Mokhtar had been forced to fund his own detention. The governor told them that authorities from the U.S. embassy would have to approve their release, or would have to come and retrieve them. Then he said his goodbyes, telling them he would see them the next day.

Mokhtar knew it was a six-hour drive to the capital, where the embassy was. No one from the embassy would make the drive to Obock. And none of it made any sense anyway, because all they had to do was get back into the boat and make the two-hour water trip to the capital.

Guards took Mokhtar and Andrew to the hotel, an array of adobe huts on a cliff above the ocean. The guards waited with them as they checked in. Once they paid, in U.S. dollars, the guards left.

The rooms were spare. Each had a cot, an end table and a fan overhead. They were not guarded. They could, if they wanted to challenge the Djiboutian authorities, try to leave the hotel compound, find the road, and try to hitchhike to the capital. Or they could take their chances at the waterfront. Find the captain. Sneak away by sea.

Neither was possible. There were too many unknowns. And the guards in their blue camouflage were likely at the waterfront. And the town was dark and desolate. There were no cars moving, no people on the streets. It was a broken-down town with a sinister air. People could disappear in a place like this.

Andrew tried his phone and found he had coverage. He called Jennifer and told her where they were. She found the number for the U.S. embassy in Djibouti City. He called, and in his most charming southern drawl, he told the woman what had happened. She promised him hospitality and safe passage once they got to Djibouti City.

They had no choice but to wait till morning.

At breakfast, there was a bizarre tableau of people in the tiny hotel restaurant. A group of mismatched army officers from North Africa. An Italian family—the parents, perhaps aid workers, whispered while their child watched cartoons on an iPad. Strangest of all was a table full of nuns, chatting amiably and seeming excited to be in Djibouti, where the temperature was now 115. Mokhtar and Andrew ate in numb silence.

After eating they took a taxi to the governor's office, where the guards told them they could take the regular commuter ferry from Obock to the capital.

"When does the ferry leave?" Mokhtar asked.

The guards weren't sure. It often left at twelve-thirty, they said. But that schedule wasn't dependable. In fact, they said, the ferry might not run at all that day.

Andrew grew agitated, and reminded Mokhtar of their timetable. They needed to be back, the beans needed to be roasted, then they needed rest. Mokhtar and Andrew came at the officials from both sides. Andrew played the demanding American, while Mokhtar spoke in more conciliatory tones. When this didn't work, they switched places and tried again. Finally it became clear that the officials couldn't countenance the two Americans getting back onto a random fishing boat and arriving in Djibouti City that way. There would be questions there, and what if they told the officials what had transpired in Obock?

Their captors were worried. They'd never stamped these two Americans in—they had no power to do so, given Obock was not an official port of entry—but at the same time, they couldn't simply let them back onto the water.

The problem was the means of passage: Mokhtar and Andrew weren't going to be allowed to get back on a boat, any boat. So they offered an alternative: they would rent a truck and hire a driver, and they would drive to the capital. The officials accepted this alternative, and soon they had secured an SUV and hired a driver, loaded their samples and were off.

It was 120 degrees, and with the humidity it felt like double that. The drive took six hours, and there was an extra man in the car— a low-level Djiboutian officer who had insisted on coming along, no doubt expecting a bribe once they got to the city.

They passed through largely uninhabited landscapes, charred red by drought and relentless heat. They drove along the Djiboutian coast, occasionally cutting inland, across parched riverbeds and rust-red valleys. All the while Mokhtar and Andrew discussed, in English, various possibilities and eventualities. They needed to be driven to the U.S. embassy, and if they made it there, the bribe, and the fact of this uninvited official tagging along, would no longer be an issue.

But they had to make it there. All along they wondered if this would be as simple as the truck taking them to the capital. What would prevent the Djiboutian official from suggesting a detour, a stopover until the matter of the bribe was worked out? There were checkpoints along the way. The Djiboutian government was trying to control the rising tide of refugees from Yemen. The first two checkpoints were easy and quick. At the third, the officer and their driver were questioned, and Andrew and Mokhtar had to present their passports. Eventually they were allowed to pass. It was late afternoon when they got to Djibouti City, a dusty city of 529,000 oppressed by the same grievous heat that kept the rest of the country under its boot heel.

Mokhtar and Andrew assumed they would be driven to the U.S.

embassy, but instead they were taken to a police station, where a young officer in stylish civilian clothes—he looked more like a model than a cop—interviewed Mokhtar and Andrew individually about why they were in Djibouti and how they got there. While Mokhtar was giving his statement, Andrew called the U.S. embassy and spoke to the woman he'd reached the night before. Carol. She said she would send someone to retrieve them.

The statements done, the stylish cop said Mokhtar and Andrew were free to go. The official from Obock did not agree. He was still waiting in the lobby. He demanded two hundred dollars, but he couldn't decide how he would define the fee. It was for his services in guiding them to the capital, he said. When Mokhtar and Andrew didn't buy it, he said it was a processing fee for their arrival in Obock. That didn't work, so he threatened to have them arrested in the police station.

"Can you help us?" Mokhtar asked the stylish cop. The stylish cop stepped in and sent the official on his way.

"I almost feel bad for him," Mokhtar said.

The embassy envoy arrived. She was a Djiboutian American from Washington, D.C., and she was so friendly and competent that Mokhtar and Andrew almost hugged her. Still, Mokhtar thought it better to have the Louisianan talk, so he stayed silent, thinking there was a remote chance he'd be detained, sent to Guantánamo.

Instead, she brought them to a travel agent and found a flight leaving the next day that would get them to the U.S. in time for the SCAA conference. It left Djibouti City at 3:00 a.m.

When they got to the airport, the customs officials were flummoxed. Mokhtar and Andrew had never been issued visas to come to

Djibouti and never had their passports stamped. Without an entry stamp, the customs officials couldn't give them an exit stamp. In a rare moment of pragmatism in the middle of the night, the customs officials decided to simply let them get on the plane. No stamps. It was as if they'd never been in Djibouti at all.

BOOK V

CHAPTER XXXIX

RETURN TO THE INFINITY

MOKHTAR'S RETURN TO THE U.S. was a circus. In San Francisco, he was met at the airport by television cameras. He did interviews with local news, NPR and Al Jazeera. He spent the night at home, with his bewildered and grateful family, and the next day, he flew to Seattle, where the Specialty Coffee Association of America conference was an unbounded success. Mokhtar gave a keynote speech, the audience stood and applauded, and he and Andrew shared a booth introducing Yemeni coffee to the specialty world. After the conference, Mokhtar was in a taxi on his way to the airport when he heard his own voice on the radio. It was an interview he'd done with the BBC.

"This guy's crazy," the driver said, not realizing the crazy man was riding in his car. When he and Andrew had hired that boat, they hadn't quite grasped how people back home would see it. They'd left Mokha during multiple firefights, hired a skiff and crossed the Red Sea—because they didn't want to miss a trade show.

Half his friends in the Bay Area thought, at least for a few hours, that he was dead. The day that Mokhtar escaped, another Bay Area Yemeni American, Jamal al-Labani, had been killed by mortar fire.

Before his name had been released, scraps of information had ricocheted around Mokhtar's network, and his friends had feared the worst.

After the conference, in San Francisco Mokhtar was met by Miriam, Justin and Giuliano, who commented on how put-together he looked for someone who just escaped a war. He was in demand from the Arab American community, from Muslim American advocacy groups, from coffee people. But finally all that died down and it was back to work. He went to Blue Bottle. James Freeman had heard his story, and now cupped his Haymah samples. Freeman scored them at 90 plus.

"How much of this can you get?" Freeman asked.

"A container. Eighteen thousand kilos," Mokhtar said.

Freeman was quiet for a moment. "That might not be enough," he said. He wanted to buy it all.

The math was preposterous. If Mokhtar could sell a container's worth of Yemeni coffee to specialty outlets, the profit margin would be significant. His farmers would be making 30 percent more than what they'd been making before.

But he needed capital. Far more capital.

He asked Ibrahim if he had ideas. Ibrahim made lists. They had tapped out everyone they knew in the Yemeni American community. They had to look elsewhere.

Mokhtar was telling Miriam about all this one day in the Mission. Miriam was just glad Mokhtar was alive—after all, she'd been the one to point him toward coffee, and he'd gone to Yemen and almost died. Now he needed money to pay for more coffee, to presumably continue going to Yemen and almost dying.

They were at Ritual Coffee Roasters, a café on Valencia Street, and were talking about all this, Yemen and its troubles and its coffee, when a woman at the next table took an interest. She was tall, blond, thin. Her name was Stephanie. "You should come to where I work," she said.

Mokhtar didn't know what coming to her work had to do with anything. But then she said she worked at a venture capital firm called Founding Fathers. It sounded intriguing.

He called Ibrahim. "Founding Fathers. That's a great name for a VC firm," Ibrahim said. But when they looked online, they found there was no VC firm called Founding Fathers. They looked up Stephanie on Facebook.

"Holy shit," Ibrahim said. "She works at Founders Fund."

Mokhtar didn't know what that meant. "Is that good?" he asked.

Ibrahim educated him. Founders Fund had made early investments in Facebook, Airbnb and Lyft. They had billions in play. Their imprimatur could make any vague notion real. They told Stephanie they would be happy to come to her work.

Then again, they thought, Founders' founder was a man named Peter Thiel, who was most recently known for his appearance at the Republican National Convention, espousing his devotion to Donald J. Trump.

"Can't worry about that yet," Mokhtar said. Founders was full of progressives, including one of its partners—a woman Stephanie thought they should meet. Her name was Cyan Banister. They looked her up. She was a noted angel investor who had bet early on SpaceX and Uber. She was also genderqueer.

"We can go," Mokhtar said. Somehow her politics, they thought, would balance out Thiel's. Then again, Thiel was gay, too. It was all very confusing.

Founders' offices were in the Presidio, a former military base on the northern shore of San Francisco, in a building renovated by George Lucas. There was a full-sized Darth Vader replica in the lobby.

"James Freeman says your coffee tastes like angels singing," Cyan said. She was very kind, and very interested and she'd done her research. There was real money in quality coffee, she knew, and she was emboldened by Blue Bottle's faith in Mokhtar. But, she said, Founders Fund couldn't be a lead investor. They didn't lead early-stage rounds.

"We have a lead investor," Ibrahim said. This was not strictly true, but Ibrahim had been thinking that if they could get a commitment from Founders Fund, they could leverage that to bring in lead funding from another VC firm, Endure Capital, based in Dubai and run by a friend of Ibrahim's, an Egyptian named Tarek Fahim.

They left the Presidio that day with an improbable plan that, a few months later, had been realized. Based on Founders' pledge, they brought in Endure. Based on Endure's commitment, they brought in funds from another firm, 500 Startups. They were suddenly a very real company. They could pay for the coffee to get to the U.S. They could pay their farmers. And they could pay themselves.

Mokhtar had a new thought. Now, he might even be able to afford his own apartment. For the time being, he was still sleeping on the floor at his parents' house in Alameda—they'd moved again—and inches from Wallead, whose snoring was an affront to propriety and the enemy of slumber.

Mokhtar had a friend who had a friend named Homera, a realtor, so he looked up her rental listings online. He laughed. He knew, immediately, that he could not afford to rent a one-bedroom apartment in San Francisco. But out of grim curiosity he scrolled through Homera's listings and stopped when he saw one in the Infinity. The pictures were astounding. Views of the Bay, all of downtown, Berkeley, Oakland, the Bay Bridge, Angel Island, Marin. In all the time he worked at the Infinity, he'd never been into one of the units. He'd delivered packages occasionally, helped residents carry things to and from the elevators, brought takeout to their doors—but he'd never been invited inside.

The price per month was farcical. He couldn't afford anything. For the hell of it he went on Craigslist. Again he saw a listing at the Infinity, this one from someone looking for a roommate to go in on a two-bedroom in the building. Thinking a shared room on short notice might be a subletter's bargain, he e-mailed the mailbox listed, saying he worked in coffee and was interested.

A reply came from someone named Shagun. He looked her up on Facebook and found she was a woman. A very attractive Indian American woman in medical school. He knew he couldn't cohabitate with her, with any woman he wasn't married to—his parents would collapse—but it couldn't hurt to meet her and see the apartment.

For the first time since he left the Infinity, he returned. He was dressed in his Rupert attire and made a point to show up a few minutes late. He didn't want to linger in the lobby, in case someone he knew was working at the desk. Shagun didn't need to know he'd been a doorman.

When she appeared, she was far too beautiful to live with, so any lingering notions he had about defying Yemeni customs were gone. Also in the lobby was a resident, an older white man, a financial manager named Jim Stauffer. Mokhtar had opened the door for Mr. Stauffer a hundred times, received and sorted his packages. Mokhtar's eyes met his, and he assumed Stauffer would come over, ask why he was back, how he was doing. Mokhtar resigned himself to being found out.

But Mr. Stauffer tilted his head, like a nearsighted man unsure what he was seeing, turned, and went on his way. Either he couldn't remember Mokhtar's name, or didn't recognize him at all.

Soon Mokhtar was alone in the elevator with Shagun, rising through the Infinity to the apartment on the twenty-third floor. She talked about medical school, about how she was looking for a roommate who worked, who was clean, who wouldn't be a distraction—she didn't say all this in so many words, but Mokhtar understood. He'd opened doors for professionals like this. He knew.

Inside, the apartment was just like the pictures in the brochures. The light was everywhere. Blue was everywhere. The city and all its glass—it was all inside the apartment. Just standing in that room would take a radical adjustment of one's equilibrium. It was like standing on the wing of a plane.

They sat down, and now she asked, gently, the questions Mokhtar assumed she'd had since shaking his hand in the lobby. How does a man your age, working in coffee, afford an apartment like this? Was he heir to some Bahraini fortune?

He told her about Yemen, about dodging bombs and Houthis to

get coffee out of the country. About his farmers, how in a few months, God willing, he'd be shipping a container full of the finest coffee in the world out of Yemen and to Oakland. And how he wanted to be in the Infinity, overlooking the Bay, when that ship come in.

"And I used to work here," he said.

"In the sales office?" she asked.

She didn't believe he'd been a doorman. He rattled off the names of half a dozen Lobby Ambassadors he knew, a few she might have met, and finally she took him at his word. He knew he couldn't live with her, with any single woman, but now there was a hunger within him. He had to live in the Infinity to prove that he could.

Two weeks later he found another listing, another sublet in Infinity B. Apparently there was a guy named Matt who lived with a guy named Jeff, who owned an analytics firm in Berkeley. Matt had to move to Ohio for work but kept the apartment and was subletting his room. His most recent lessee was a Russian business student who was moving out.

The price for the one-bedroom sublet was more than Mokhtar had ever spent on anything in the United States. But he had a vision. The vision featured him, Mokhtar Alkhanshali, living on the thirtieth floor or whatever floor Jeff and Matt lived on, standing on a balcony with everyone he loved, watching his coffee come into port.

Mokhtar called Matt, and Matt thought Mokhtar seemed a worthy successor to the Russian oligarch. All Mokhtar had to do was meet Jeff, who would have final say. Mokhtar walked through the lobby, not recognizing anyone working the front desk, and got into the elevator. On the thirty-third floor, Jeff opened the door. A tall white man in his forties, he offered Mokhtar wine. Mokhtar declined. Jeff poured himself a glass and they talked about the Russian, about their work schedules, and all the while Jeff seemed to want to ask the same question Shagun had implied. Are you some kind of Saudi prince? When Mokhtar had said he was in coffee, Jeff assumed he was a barista. Eventually Mokhtar noticed a high-end hand grinder on the counter, and found a way to work Blue Bottle into the conversation. Then it was over. Jeff went to Blue Bottle every day. He offered Mokhtar the sublet, and Mokhtar, defying all fiscal responsibility, took it.

Stephen offered help with the move, but there wasn't much to contend with. Mokhtar had one suitcase and two garbage bags. They parked around the corner and carried Mokhtar's worldly possessions into the lobby of the Infinity Tower B.

A young man, Jonathan, was working the front desk that day. The phone was ringing and the lobby was frantic. Jonathan was supposed to give Mokhtar the key Jeff had left for him, but he couldn't find it. Mokhtar and Stephen waited on the lobby's leather couches.

"You okay?" Stephen asked.

"I'm fine. Why?" Mokhtar said.

"You keep getting up to open the door for people," Stephen said.

Mokhtar had been up and down half a dozen times. He couldn't help it. "Sorry," he said. "Force of habit."

"You don't work here anymore," Stephen said.

"I know. I know," Mokhtar said.

A week later, Mokhtar was driving over the Bay Bridge, heading to San Francisco. The city was as bright as a chandelier. His father was in the front seat, his mother in back. They'd just eaten dinner out, Mokhtar's treat. "Remember the Infinity, where I used to work?" he asked them.

They remembered.

"They're having an open house tonight," Mokhtar said. "You want to go?"

His parents had two sons who'd worked at the Infinity, so seeing inside one of the towers wasn't without interest. But it was eight o'clock on a weekday. Why would the building be holding an open house now?

There were no other visitors in the lobby. No signs of an open house. Mokhtar hoped only that his parents would believe a little longer. In the elevator he pushed thirty-three. He'd arranged it so Jeff would be gone when they arrived. He slipped the key in the slot and opened the door. As always the city was alive in every window, the candelabra of downtown reflected on the apartment's polished floors.

"There's no one here," his mother said.

Mokhtar brought his mother to the balcony and they breathed in the air from the Bay, from the sky at that height, but his father was still near the doorway.

"He's afraid of heights," his mother said. "You didn't know?"

Mokhtar didn't know. They'd never been higher than the third floor, in their old apartment in the Tenderloin.

Mokhtar brought his mother back inside, and by then his father had seen the photographs. Mokhtar had arranged framed photos of his mother and father on the coffee table.

"Why are those here?" his mother asked.

"Mom, Dad, sit down," Mokhtar said.

They sat down.

"I'm doing really well now," he said. "I'm working hard and the company's doing well. I want you to be proud of me, and I want to provide for you." He told them about his coffee, about the orders, about the ship coming.

"That's so good, Mokhtar," his mother said. "But I still don't understand why those pictures of us are here."

"Mom," Mokhtar said, "because this is where I live."

CHAPTER XL

COFFEE ON THE WATER

CONDITIONS IN YEMEN WERE deteriorating. Virtually no goods were being shipped out of the country. Activity at the ports was concentrated on importing essentials. Medicine was scarce and the vast majority of the country was suffering from food insecurity. The UN considered Yemen on the brink of famine. No one was prioritizing the export of coffee to international specialty roasters.

But Mokhtar kept his farmers harvesting. The Widow Warda, the General, and the rest of the farmers of Haymah continued their work—their region was nearly untouched by the war—and they continued to ship the red cherries to Mokhtar's warehouse in Sana'a. His sorters came to work every day. With the airstrikes, any power they could get had to come from their own diesel-fed generators.

Mokhtar checked in every morning, at 4:00 a.m. California time. He talked with Andrew and Ali to make sure everyone was safe, and he dealt with financial issues and logistics. The problems came from all sides. One day a sorter was unable to make it to work when a Saudi bomb had cratered the road she needed to take to get to work.

Another sorter's husband was being pressured to fight for the Houthis; they'd had to go into hiding.

Then there was the matter of the GrainPro bags. That Mokhtar needed the bags at all was good news. They had enough green beans to fill a container. But to transport them by sea, it wasn't enough to put them in the traditional burlap sacks. If Mokhtar was to send the message that his coffee would be different and superior, it had to start with the packaging, with ensuring that the coffee arrived without the odor of the sea, of the ship's hold, whatever else it was carrying or had carried.

GrainPro bags were the industry standard—thick plastic bags that kept moisture in and interfering elements out. In the U.S., or virtually any other part of the world, getting hold of GrainPro bags would be as easy as a phone call and a visit by UPS. But to get these bags to Yemen, during a war, was beyond the realm of reason.

Mokhtar managed to get twelve hundred bags shipped to Ethiopia. That took two weeks. But no one in Ethiopia could get them to Yemen. He made calls to Djibouti and found a freighter that was making trips between Djibouti City and Mokha. That took another six weeks. In all, it was two months before he could get the bags to Sana'a. There, the sorted coffee was placed in carefully labeled bags. Sealed, the bags were brought to the port of Aden, and were ready to be shipped to Oakland.

It was difficult sometimes to see all this as essential. People were dying in Yemen. The country was collapsing, and in his San Francisco high-rise, Mokhtar was waking up every morning at 4:00 a.m. to call Sana'a about coffee. About when the container could leave Yemen.

But there was a lot of money at stake. Omar's money. His investors' money. Hubayshi's money. There was the faith of all his farmers to think about, about their pickers and his processors. And now Mokhtar had his own staff in San Francisco. He'd hired his old friend Ibrahim Ahmed Ibrahim as CFO. (Ibrahim's wife, Salwa, was being as supportive as she could, given they had a fifteen-month-old baby and her husband had just quit his well-paying job at Intuit to work with Mokhtar, whose most applicable experience was selling shirts and Hondas.) Mokhtar had brought on Jodi and Marlee from Boot as directors of quality control, and his old friend Jeremy as his executive assistant.

None of this could be sustained unless the coffee left the port. Stephen and Mokhtar talked a dozen times a day, from their respective apartments. They had been working with a shipping company, Atlas, and the company's owner, Craig Holt, had taken a personal interest in Mokhtar's mission. For months he'd been wrestling with how to get the coffee out of Yemen. One day in late December, Holt sent word that they'd be picking up Mokhtar's container on New Year's Eve.

On January 1, 2016, the coffee was on the water. The ship was called the MSC *Rebecca*.

CHAPTER XLI

THE *LUCIANA*

IN JEDDAH, MOKHTAR'S CONTAINER was transferred from the MSC *Rebecca* to a larger ship, the MSC *Luciana*, and the *Luciana* made its way from Jeddah to Singapore to Long Beach. The trip took almost two months. Finally, late in February, Holt told Mokhtar that his coffee had arrived in the United States and was going through customs in Long Beach. His best estimate was that the container would be in Oakland on Saturday, February 25.

Mokhtar called his parents and Stephen and Ibrahim, and texted Miriam and Ghassan, Giuliano and Justin. He texted everyone. *Come see my coffee come into port,* he told them. He planned to have a party on the balcony of the Infinity, with everyone there, everyone he loved, to watch his ship come in. He needed sparkling cider. Nonalcoholic champagne. Soda, cheese, crackers, dip. He had to go to the store.

Then again, there were no guarantees the coffee would arrive on Saturday at all. There was no telling how long a ship would wait, how long any given container would be searched—especially a container originating in Yemen.

Mokhtar went to bed on Thursday and woke up to call Yemen.

That was still his routine, calling at three or four in the morning, which was afternoon in Yemen, to check in with Ali and Nurideen. The call that morning was full of trouble. The peace talks were going nowhere. The processing plant had no electricity and the women were worried for their jobs. Mokhtar was worried, too. The next harvest wasn't for months and there wasn't much for the women to do. His investors had strongly urged him to lay the women off; it made no sense to keep so many sorters on the payroll when there were no beans to sort.

But if he laid them off, they'd never get other jobs, not in the middle of a war, and how would Mokhtar find and train another team of sorters for the next harvest? So he kept them—Baghdad, Samera, Ruqayah, Shams, Alham and Alham (there were two Alhams)—he kept them all on salary. They were grateful, especially given most of their husbands were unemployed in Sana'a, a city under attack.

Mokhtar woke up late Friday with a head full of dark thoughts, with the unshakable sense that everything was about to go wrong. That the container would be detained. That the beans would arrive spoiled. That he would be buried in debt.

Stephen was at a wedding in Florida, but he was tracking the latest information about the *Luciana*. Mokhtar shot him a text, triple-checking the ship's status. Was it still coming the next day?

Seconds later, Stephen called. "It's coming now."

"What is?" Mokhtar asked. He sat up in bed.

"The ship," Stephen said. "It'll be in Oakland in two hours."

Stephen could see the ship's progress on his phone. The MSC *Luciana* was steadily clicking up the coast.

"It can't be. You sure?" Mokhtar asked.

Stephen got off the phone to check with Atlas. Atlas said the latest information indicated the ship would arrive that night at 10:00 p.m. But the tracking app on Stephen's phone showed the ship speeding up the Pacific coast, approaching the Golden Gate.

Mokhtar got out of bed. He ran around the apartment. He had no idea what to do. The ship was early. It was scheduled to arrive at two and it was already noon.

He called his mom and got her voice mail. His dad was driving his bus. He called Miriam, who was down on the peninsula, an hour away.

Ibrahim was in meetings in San Francisco, wrapping up his last day with Intuit. The only person Mokhtar could reach, and who could make it to the Infinity in time, was the guy writing a book about all this. It was not the way Mokhtar pictured it.

ON HIS PHONE, Mokhtar watched the little icon that was the MSC *Luciana* chugging up the coast, one pixel at a time. Past Monterey. Past Pacifica. Mokhtar went to the balcony, thinking he might be able to see the ship. Nothing. It hadn't passed through the Golden Gate yet.

The door buzzed. The writer was there, and we stood there, panting, laughing at this, the fact that this was really happening. But there was no nonalcoholic champagne or cider. There were no close friends, no family. It was just the two of us, and the ship was so close.

Mokhtar watched the app on his phone.

"Look. It's going under the Golden Gate Bridge," he said.

There it was, the video-game version of the ship, a tiny red arrow

314

on his little screen. Again and again we looked up from the phone and then north, through the city, as if we could see the ship through the hills and buildings blocking the view of the Bay.

We realized where we'd see it first. There were two buildings, part of 1 Market Plaza, arranged at a diagonal just two blocks north of the Infinity Towers. There was a small gap between them, revealing a small stripe of the Bay's cobalt blue where the *Luciana* would pass.

The sun was high and white. The day was impossibly bright. There were a few sailboats out, a ferry or two, nothing else. No ships. No tankers. Whenever a ship appeared between those towers, it would be the *Luciana*. There was nothing else like it on the water.

Below, we could see all of Treasure Island, its low-slung white buildings. "It's going to go right past my old house," Mokhtar said.

He checked his phone again. The *Luciana*'s red arrow passed Fisherman's Wharf and was rounding North Beach and the Embarcadero. The real *Luciana,* we were sure, would be visible any second.

And there it was. Between the towers, the black nose of the ship.

"Oh my God," Mokhtar said.

The *Luciana*. It said so right on the bow. The ship was stacked high with containers of white and blue, yellow and green.

Mokhtar turned on his camera and narrated. "We're here on the twenty-sixth of February. Between those two buildings, right over there, that ship is carrying eighteen thousand kilograms of the world's best coffee. From Yemen."

The ship was passing Treasure Island, and the Ferry Building, an American flag high above its tower, seagulls circling. Tugboats, two or three or four, guided the *Luciana* through the Bay. Mokhtar's phone dinged. It was Ibrahim. He'd left work early and was on his way.

"You need to come now," Mokhtar told him. "Double-park the car. Anything."

Minutes later Ibrahim was there, too, on the balcony. He and Mokhtar hugged. The *Luciana* was still passing Treasure Island. Mokhtar called Stephen. He picked up. Stephen's grinning face, redder from the Florida sun, filled the screen. There were palm trees behind him.

"You see that?" Mokhtar said. "It's the *Luciana*! It's right there!"

Stephen tilted the phone to reveal a young woman next to him. "This is Leigh. She's getting married tomorrow."

"Congratulations, Leigh," Mokhtar said. "I wish you the best in this life. It's going to be wonderful." Everything seemed wonderful.

"Aw man, I wish I was there," Stephen said.

"You are," Mokhtar said. "You *are* there!"

He pointed his phone's camera to the ship so Stephen could watch. They hung up. He had other people to call. Miriam. He reached her, showed her the *Luciana* passing steadily.

"Remember the text message you sent me?" he asked.

You ever look across the street?

She remembered.

"But I can't FaceTime," she said. "I'm in sweatpants."

Mokhtar called his mom. He stood on the edge of the balcony, the ship over his shoulder, the water and Treasure Island just beyond.

"I love you," he told her, and kissed the phone.

And soon the ship was out of view.

CHAPTER XLII

THE DOORMEN UNITE
AND OPEN THE ROOF

"WE SHOULD GO TO the roof," Mokhtar said. "We can see everything from the roof."

He led Ibrahim and me downstairs, where a young Lobby Ambassador named Nick was standing behind the front desk. When we all appeared, rushing toward the desk, Nick's eyes widened as if he expected to be overrun. But Mokhtar knew Nick. Mokhtar had had him over for dinner in his apartment. Nick was from Oakland, had been working at the Infinity for seven months. The front desk was, for him, a stopgap on the way to what he hoped would be a job in finance.

Now Mokhtar was asking him to break a very clear-cut rule against letting anyone, resident or not, onto the roof. The Infinity roof wasn't built for entertaining. It was an industrial roof, full of HVACs and cables. There were no adequate railings, nothing at all to accommodate anyone.

Then again, this was Mokhtar. And there was this ship, Mokhtar was explaining, and it would only come in once, and—

"Okay," Nick said. "Fine."

He led us into the elevator and to the thirty-fifth floor. He found a nondescript door, and used his master key to open it, sighing all the time. He led us up two flights of stairs and opened another door.

We were on the roof. It was dizzying and too bright. The view was almost unobstructed. Mokhtar promised Nick we'd be only a minute, that he'd never tell a soul.

Nick looked worried. He'd not only let a resident and two nonresidents onto the roof, which was not at all ready for visitors, but he'd abandoned his post at the front desk.

"Gotta get back," he said, and disappeared into the building.

We could still see the ship. It was still chugging toward the Port of Oakland. From the roof, amid the churning machinery that kept the Infinity the right temperature year-round, we saw all of the Bay Bridge, the cars tiny, the trucks so small. We could see the tankers waiting in the South Bay, we saw all of San Francisco, all of Treasure Island.

Mokhtar couldn't stop laughing. Then he cried a little. Then he laughed some more. Ibrahim laughed, too. He'd just had his last meeting for Intuit and now he was on the roof of the Infinity, watching their coffee come into port.

"Look," Mokhtar said, and pointed down. "You can see the courtyard. The one with the monk." Ibrahim and I looked and saw, thirty-seven floors down, the corner of the old Hills Bros. courtyard, but the monk was out of view. And soon the *Luciana* was out of view, too. It had ducked behind Infinity Tower D, the easternmost Infinity, five stories taller than Infinity B and blocking the view of the shipyard.

"We have to get over there," Mokhtar said.

Ibrahim and I insisted that it didn't matter, we'd seen the ship, there was no reason to go all the way down and up again, to another roof just to see a bit more.

"It'll be easy," Mokhtar said, and he led us down from the roof, into the elevator and down the thirty-five floors, through the lobby, where we thanked Nick and Mokhtar asked if he could let us onto the roof of Tower D.

"Ask Ana," Nick said, more anxious than before. "She's working D."

Mokhtar knew Ana. Her full name was Borana Haxhija and she was like a sister. Her parents had fled from Albania, had come to the U.S. around the same time that Mokhtar's parents came from Yemen. The Haxhijas had settled in the Tenderloin, too, just a few blocks away from where the Alkhanshalis lived between two porn stores. Ana had gone to Galileo High School and now was working two jobs while finishing her degree at San Francisco State. When Mokhtar burst into the lobby and saw her, he knew she'd allow it.

"Can we get up onto the roof?" he asked.

She didn't ask why. She just handed him the key.

We rode up to the forty-second floor, took the stairs to the roof, and the view was clear for fifty miles in every direction. We could see the *Luciana* turning into the Port of Oakland. And below, we could see the courtyard, all of it, and the monk, holding the cup to the sky.

EPILOGUE

ON JUNE 9, 2016, PORT of Mokha coffee was made available for the first time at Blue Bottle coffee shops around the United States. It was the most expensive coffee Blue Bottle had ever sold. Complete with a cardamom cookie made from Mokhtar's mother's recipe, it cost $16 a cup.

Willem and Jodi and Marlee—everyone at Boot—celebrated. Tadesse Meskela texted his congratulations from Addis Ababa. Mokhtar heard from Camilo Sánchez in New York, from Graciano Cruz in Panama. He heard from his farmers in Yemen. *Wherever you go, we go with you,* they said. Mokhtar had by then paid for weddings, for funerals, medical procedures and college tuitions.

Word of Mokhtar's work had already spread throughout Yemen. Farmers across the country were bringing their coffee to Port of Mokha in Sana'a, hoping to engage in direct trade and raise prices for their crop. As of spring of 2017, farmers in Haymah had replaced seventeen thousand qat plants with coffee trees.

By July 2017, Port of Mokha coffee was available (at more affordable price levels, too) all over North America, Japan, Paris and Brazil. Across four continents, the coffee in the first container sold out in forty-five days. Port of Mokha's second shipment, airlifted from Yemen to Jordan to San Francisco, arrived on January 5, 2017. It sold out in thirty-two days.

In February 2017, the *Coffee Review* awarded Port of Mokha's Haymah microlot a 97 rating, the highest score issued in the publication's twenty-year history.

Andrew Nicholson returned to Yemen on a cattle boat one month after the Seattle coffee conference. Eventually he brought his family back to the U.S., but continued to travel to Yemen to operate Rayyan until he was kidnapped in Sana'a by a rebel group. He was held for one month, unhurt, and upon release rejoined his family in the U.S. The Rayyan mill is still operating and exporting coffee all over the world.

Mokhtar moved out of the Infinity in 2016. His apartment was too lofty and too lonely. He'd only wanted to be there to watch his coffee come into Oakland. That's where he moved when he left the Infinity, to Oakland, not far from the Fruitvale BART station. This is where Port of Mokha has its coffee lab and where his varietals are stored, cupped and roasted.

Given the troubles in Yemen, Mokhtar's grandfather Hamood had returned to the United States, staying with relatives in California's Central Valley. Mokhtar went to visit him one day. As he approached the house, he saw his grandfather sitting outside alone, his head resting on his cane. Mokhtar approached him, kissed his knees and hands and forehead. It was Eid, and as was customary, Mokhtar had brought Hamood a gift. It was an envelope filled with hundred-dollar bills.

"Where'd you get this?" Hamood asked.

"This is from the boy worth less than a donkey," Mokhtar said, and smiled.

Mokhtar had never seen his grandfather cry. He sat next to him and held him.

He stayed the rest of the day with his family, and drove back to Oakland that night. He had a few hours to sleep before Yemen would be calling.

ACKNOWLEDGMENTS

This book benefited from the generosity of countless people who shared their memories and expertise and in the process made this book as accurate and thorough as possible. Thank you to: Miriam Zouzounis, Jeremy Stern, Giuliano Sarinelli, Benish Sarinelli, Justin Chen, Ibrahim Abram Ibrahim, Andrew Nicholson, Ghassan Toukan, Summer Nasser, Willem Boot, Catherine Cadloni, Jodi Wieser, Marlee Benefield, Stephen Ezell, James Freeman, Wallead Alkhanshali, Faisal Alkhanshali, and Bushra Alkhanshali, Nasrina Bargzaie, Zahra Billoo, Temesgen Woldezion, Shaimaa al-Mukhtar, Maytha Alhassan, Marwa Helal and Aben Ezer. Daniel Gumbiner was indispensable in bringing this book to fruition; his close readings and tireless dedication to accuracy were invaluable. Em-J Staples provided stalwart support, research and encouraging words over twenty-eight months. Profound gratitude is owed to Peter Salisbury, scholar of Yemen, and to Meghan O'Sullivan, formerly of the U.S. State Department, who sat down with me in 2015 to talk about the situation in Yemen vis-à-vis Saudi Arabia, Iran and the interests of the United States and its allies. Fatima Abo Alasrar, a keen-eyed Yemeni American journalist, provided acute analysis and background, as did the poet-scholar Steven C. Caton, professor at Harvard. Thank you to the Asian Law Caucus of San Francisco. Thank you to the Council on American-Islamic Relations. Thank you to our friends and guides in Yemen, Djibouti and Ethiopia. Thank you to Jay and Kristen Ruskey. For their close readings of the book and various other forms of assistance, thank you Tish Scola, Paul Scola, Amanda Uhle, Inder Comar, Ebby Amir, Becky Wilson and Kevin Feeney. Tom Jenks did a wonderful job getting the book down to fighting weight; thank you, sir. Jennifer Jackson has been my editor at Knopf for almost sixteen years now, and my gratitude for our long and happy relationship cannot be overstated; to be bathed in the light of her enthusiasm over so many years makes a human feel blessed and strong. Sonny Mehta has supported the work Jenny and I have done all this time and his benevolence has made all things possible. Thank you to all at Knopf, especially Andy, Paul, and Zakiya. Andrew Wylie has been a constant friend and champion for almost twenty years; I feel fortunate to be his client and to benefit from his, and his staff's, unwavering care and curation.

The following extraordinary books were crucial in understanding the history of coffee and the business of coffee: *Uncommon Grounds: The History of Coffee and How It Transformed Our World,* by Mark Pendergrast; *The Joy of Coffee,* by Corby Kummer; *Coffee: A Dark History,* by Antony Wild; *Javatrekker: Dispatches from the World of Fair Trade Coffee,* by Dean Cycon; *From These Hands: A Journey Along the Coffee Trail,* by Steve McCurry; *The Devil's Cup: A History of the World According to Coffee,* by Stewart Lee Allen. Willem Boot and Camilo Sánchez wrote an essential overview of the coffee trade of Yemen in their paper *Rediscovering Coffee in Yemen: Updating the Coffee Value Chain and a Marketing Strategy to Re-Position Yemen in the International Coffee Markets.* The following books about Yemen were enormously helpful: *Yemen Chronicle: An Anthropology of War and Mediation,* by Steven C. Caton; *The Merchant Houses of Mocha: Trade and Architecture in an Indian Ocean Port,* by Nancy Um; *Tribes and Politics in*

Yemen: A History of the Houthi Conflict, by Marieke Brandt; *Yemen: The Unknown Arabia,* by Tim Mackintosh-Smith; *The Last Refuge: Yemen, al-Qaeda, and America's War in Arabia,* by Gregory D. Johnsen. For an excellent primer on the Tenderloin neighborhood of San Francisco, please find *The Tenderloin,* by Randy Shaw. A compelling and concise survey of the coffee trade is presented in a documentary called *Black Gold: Wake Up and Smell the Coffee,* directed by Marc Francis and Nick Francis. *The Fight for Yemen,* a documentary produced by PBS and Frontline and directed by Safa Al Ahmad, is a cogent and perceptive overview of the rise of the Houthis.

For a more detailed bibliography and list of sources, please see www.daveeggers.net/monkofmokha.

This book would not have been possible without the vision and enthusiasm of Wajahat Ali. Thank you, my friend.

This book would not have been possible without the boundless honesty and courage of Mokhtar Alkhanshali. Thank you, my brother.

Nothing would be possible without VV. Thank you, my love.

THE MOKHA FOUNDATION

The author and Mokhtar Alkhanshali have set aside proceeds from this book to establish the Mokha Foundation, which directly invests in improving the quality of life in Yemen in a variety of ways that include supporting farmers and their families, preserving natural resources, and disrupting the refugee crisis at the front line. To join in this effort, visit www.mokhafoundation.org.

826 VALENCIA IN THE TENDERLOIN

826 Valencia, a writing and tutoring center based in San Francisco, opened a second location at the corner of Golden Gate and Leavenworth, in the heart of the Tenderloin district. Fronted by a retail shop, King Carl's Emporium, which sells sea-exploration gear for humans and fish, 826's Tenderloin center offers free writing workshops and an array of other services to young people in the neighborhood and throughout the city. It is a safe and welcoming place. For more information, see www.826valencia.org.